ACCESS TO CHEMISTRY

Periodic Table of the Elements

RS•C
ROYAL SOCIETY OF CHEMISTRY

Group 1	Group 2		3	4	5	6	7	0
1 H 1.008								2 He 4.003
3 Li 6.939	4 Be 9.012		5 B 10.81	6 C 12.011	7 N 14.007	8 O 15.999	9 F 18.998	10 Ne 20.180
11 Na 22.990	12 Mg 24.305		13 Al 26.982	14 Si 28.086	15 P 30.974	16 S 32.066	17 Cl 35.453	18 Ar 39.948

Transition metals:

21 Sc 44.956	22 Ti 47.88	23 V 50.942	24 Cr 51.996	25 Mn 54.938	26 Fe 55.847	27 Co 58.933	28 Ni 58.69	29 Cu 63.546	30 Zn 65.39
39 Y 88.906	40 Zr 91.224	41 Nb 92.906	42 Mo 95.94	43 Tc 99*	44 Ru 101.07	45 Rh 102.91	46 Pd 106.42	47 Ag 107.87	48 Cd 112.41
57 La 138.91	72 Hf 178.49	73 Ta 180.95	74 W 183.85	75 Re 186.20	76 Os 190.20	77 Ir 192.20	78 Pt 195.08	79 Au 196.97	80 Hg 200.59
89 Ac 227*	104 Rf 261.11	105 Db 262.11	106 Sg 263.12	107 Bh 262.12	108 Hs 265*	109 Mt 266			

Groups 19 K 39.098 / 20 Ca 40.078 / 37 Rb 85.47 / 38 Sr 87.62 / 55 Cs 132.91 / 56 Ba 137.33 / 87 Fr 223* / 88 Ra 226.02

31 Ga 69.723	32 Ge 72.61	33 As 74.923	34 Se 78.96	35 Br 79.904	36 Kr 83.80
49 In 114.82	50 Sn 118.71	51 Sb 121.75	52 Te 127.60	53 I 126.90	54 Xe 131.29
81 Tl 204.38	82 Pb 207.20	83 Bi 208.98	84 Po 209*	85 At 210*	86 Rn 222*

Lanthanum series

58 Ce 140.12	59 Pr 140.91	60 Nd 144.24	61 Pm 145*	62 Sm 150.36	63 Eu 151.97	64 Gd 157.25	65 Tb 158.92	66 Dy 162.50	67 Ho 164.93	68 Er 167.26	69 Tm 168.93	70 Yb 173.04	71 Lu 174.97

Actinium series

90 Th 232.04	91 Pa 231.04	92 U 238.04	93 Np 237.05	94 Pu 244*	95 Am 243*	96 Cm 247*	97 Bk 247*	98 Cf 251*	99 Es 254*	100 Fm 257*	101 Md 258*	102 No 259*	103 Lr 260*

* radioactive

For more information see the Visual Elements Periodic Table at http://www.chemsoc.org/viselements

Access to Chemistry

A. V. JONES
M. CLEMMET
A. HIGTON
E. GOLDING
Nottingham Trent University, Nottingham, UK

RS•C
ROYAL SOCIETY OF CHEMISTRY

ISBN 0-85404-564-3

A catalogue record for this book is available from the British Library

Published by The Royal Society of Chemistry,
Thomas Graham House, Science Park, Milton Road, Cambridge CB4 0WF, UK

For further information see our web site at www.rsc.org

Typeset by Ward Partnership
Printed and bound by Polestar Wheatons Ltd

Preface

It is assumed that you have picked up this text because you have the motivation to want to 'know about' chemistry. This might be because you want to go on and study chemistry further, or it could be that a knowledge of chemistry is needed to support another subject, *e.g.* biology or environmental studies, health studies, sports science or engineering courses. This text will be suitable for you. The course generally assumes a starting point of a 'pass' in GCSE single science or equivalent. Many of the units start at such a level that even people who have long since forgotten the science they once knew should be comfortable.

It could be that in the past you have found chemistry uninteresting and a little 'difficult' but now see that a knowledge of chemistry is needed to prepare you for university courses or for a balanced education. Alternatively, you might want a basic understanding of chemistry to give you a better insight into areas such as health, home chemicals, business market trends, gardening, *etc*. This text can help you get to grips with an understanding of chemical molecules, their structures and uses. They make up EVERYTHING, including the authors as we write this and you as you read it.

One of the great things about self-motivation, as compared with enforced study, is that you can work at your own pace; BUT, work is still necessary for understanding. There is no such thing as 'instant understanding' of chemistry or any other subject or skill. How many times did you fall from a bike or skateboard before you mastered it? Have you ever forgotten how to ride a bike? You stuck at it because YOU wanted to master the task. The study in this book is very similar: if you want to succeed, it requires 'stick-ability'. In the following sections we will give a few hints to help you study the material most effectively. We must make the best use of study time, and so we have given some hints on approaches that other students have found helpful. Read these before you start intensive study. The principles apply to all subjects, not just chemistry.

The text will also be of value to those studying 'A' level or GNVQ chemistry as it gives opportunities for self-study. It has also been used by technical staff in laboratories as a source of self-study.

IF YOU REALLY WANT TO SUCCEED . . . THEN GO FOR IT!

Contents

Aims   xvii
Format of Materials   xviii
Study Guide   xix

Module 1   Building Blocks of Matter

Unit 1.1   Particles and Properties   3

1.1.1   Properties   5
1.1.2   Change of State   5
1.1.3   Particles   7
          State and change of state   7
1.1.4   Evaporation   9
1.1.5   Diffusion   10
1.1.6   Dissolving   11
1.1.7   Pressure   11

Answers to diagnostic test   14
Further questions   15

Unit 1.2   Inside the Atom   17

1.2.1   Elements   18
1.2.2   Compounds   19
1.2.3   Atoms   20
1.2.4   Smaller Particles   20
1.2.5   Electronic Configuration   21
1.2.6   Further Reading   22

Answers to diagnostic test   23
Further questions   23

Unit 1.3   Atoms, Elements and Order   25

1.3.1   Three Elementary Particles   26
1.3.2   Symbols   26
1.3.3   Atomic and Mass Numbers   27
1.3.4   Isotopes and Relative Atomic Mass   27
1.3.5   Periodicity   28

Answers to diagnostic test   31
Further questions   31

Unit 1.4   The Nucleus of the Atom   33

1.4.1   The Discovery of Radioactivity   34
1.4.2   Types of Radiation   34

1.4.3   Transmutation of Elements   35
1.4.4   Half-life   36

Answers to diagnostic test   39
Further questions   39

Unit 1.5   Electrons   41

1.5.1   Electrons in Atoms   42
1.5.2   Absorption Spectra   42
1.5.3   Why Do Elements Only Absorb Certain Wavelengths?   43
1.5.4   More Accurate Spectroscopes, and Quantum Numbers   44

Answers to diagnostic test   47
Further questions   47

Module 2   Chemical Bonding in Materials

Unit 2.1   Atoms and Ions   51

2.1.1   Why Atoms Combine   52
2.1.2   How Atoms Combine   52

Answers to diagnostic test   56
Further questions   58

Unit 2.2   Molecules and Sharing Electrons   59

2.2.1   Sharing Electrons   60
2.2.2   Bonds   60
2.2.3   Equal and Unequal Shares   61
2.2.4   Shapes of Orbitals and Bonds   62

Answers to diagnostic test   64
Further questions   65

Unit 2.3   Bonding and Properties   67

2.3.1   Lattices   68
2.3.2   Giant Covalent Structures   68
        Summary   69
2.3.3   Giant Metallic Structures   69
2.3.4   Different Bonding, Different Properties   70
        Melting and boiling points   70
        Electrical conductivity   71
        Solubility in water   71
2.3.5   Hydrogen bonding   72

Answers to diagnostic test   73
Further questions   73

Unit 2.4   Electronic Configuration and Reactivity   75

2.4.1   Loss of Electrons   76
2.4.2   Gain of Electrons   77
2.4.3   Trends in the Periodic Table   78
2.4.4   Reactivity Series   78

Answers to diagnostic test   82
Further questions   82

Unit 2.5   Oxidation Numbers   83

2.5.1   Oxidation Numbers and Group   84
2.5.2   Oxidation Numbers and Formulae   85
2.5.3   The Transition Metals   85
2.5.4   Multi-atom Ions   88
2.5.5   Summary of Oxidation and Oxidation Numbers   89

Answers to diagnostic test   90
Further questions   90

Module 3   Different Types of Chemical Reaction

Unit 3.1   Synthesis and Combining   93

3.1.1   Binary Compounds by Joining Elements Together   94
3.1.2   Fuels Combining with Oxygen   95
3.1.3   Organic Synthesis   95

Answers to diagnostic test   96
Further questions   96

Unit 3.2   Decomposition – The Breaking Up of Compounds   97

3.2.1   Thermal Decomposition   99
3.2.2   Electrolysis of Molten Compounds   101
3.2.3   Electrolysis of Aqueous Solutions   102

Answers to diagnostic test   104
Further questions   104

Unit 3.3   Exchange Chemical Reactions   105

3.3.1   Displacement   106
3.3.2   Acids and Metals   107
3.3.3   Ion Exchange Reactions   108
3.3.4   Neutralisation of Acids   109
3.3.5   Summary of Chemical Reactions of Acids and Alkalis   110

Answers to diagnostic test   111
Further questions   111

**Unit 3.4   Some Further General Characteristics of Chemical
                Reactions**   113

3.4.1   Heat Energy   114
3.4.2   Electricity   114
3.4.3   Redox Reactions   116

Answers to diagnostic test   118
Further questions   118

Module 4   Quantitative Aspects of Chemical Reactions

Unit 4.1   Mathematical Concepts   121

4.1.1   Introduction   122
           General points   122

4.1.2	Standard Form (Exponential) Numbers and Powers of 10	122
	Rules for working with exponential numbers or numbers in standard form	123
4.1.3	Calculations with Numbers in Standard Form	124
	Adding and subtracting	124
	Multiplication and division	125
4.1.4	Using a Calculator	125
	Work with numbers written in standard form	125
	Work with logarithms	126
	Significant figures, decimal places, and 'rounding up'	127
4.1.5	Rearranging Equations	127
	Treat both sides of the equation the same way	128
	Cancelling out	129
4.1.6	Graphs	129
	Plotting a graph	129
	Drawing the line – the best fit	130
	The gradient or slope	130
	Using the graph	131
	The exponential graph	131
4.1.7	Units	132
	Answers to diagnostic test	134
	Further questions	134
4.1.8	Extension – Theory of Logarithms	135
	Logarithms to base 10 (abbreviated to $\log_{10}$ or, simply, log)	135
	Logarithms to the base e (natural logs, abbreviated to $\ln_e$)	136
Unit 4.2	**Chemical Laws – The Law of Constant Composition and the Law of Conservation of Mass**	137
4.2.1	The Law of Constant Composition	139
4.2.2	The Law of Conservation of Mass	139
4.2.3	Balancing Equations	141
4.2.4	Rules for Balancing Equations	142
	Further hints to help you	143
	Answers to diagnostic test	147
	Further questions	147
Unit 4.3	**Reacting Quantities and the Mole**	149
4.3.1	Introduction to the Mole	151
	For molecules and ionic compounds	152
4.3.2	Using Moles	153
	Writing a chemical recipe	153
	Working out a product yield	154
4.3.3	Percentage Yields	155
4.3.4	Molar Excess and Limiting Quantities	156
	Questions	157
	Answers	157
4.3.5	Percentage Composition, and using the Law of Constant Composition to find the Empirical and Chemical Formula of a Compound	158
	Difference between the chemical formula and the empirical formula	159

Reverse calculations; percentage composition
      from formula   160

Answers to diagnostic test   161
Further questions   161

Unit 4.4   Strengths of Solutions   163

4.4.1   Units of Volume and Concentration   164
         Units of volume   164
         Concentration   164
4.4.2   Diluting Solutions   167
4.4.3   Saturated Solutions   168
         Concentrated acids   169
         Diluting concentrated acids   169
4.4.4   Other Units for Concentration   169
         Parts per million (ppm)   170
         Parts per billion (ppb)   170
         Percent by mass   170
         Percent by volume   171

Answers to diagnostic test   172
Further questions   172

Unit 4.5   Gases   173

4.5.1   Introduction   175
4.5.2   Units   176
         Pressure   176
         Volume   177
         Temperature   177
         Standard temperature and pressure (STP)   177
         Molar volume   177
4.5.3   Kinetic Theory   178
4.5.4   Ideal Gases and the Gas Laws   178
         Boyle's Law: the effect of pressure on the volume
            of a gas at constant temperature   179
         Charles's Law: the effect of temperature on the
            volume of a gas at constant pressure   179
         The effect of temperature on the pressure of a gas
            at constant volume   180
         The Combined Gas Law   180
4.5.5   The Ideal Gas Equation – Calculating the Mass and
            Relative Molecular Mass of a Gas   182
4.5.6   Changing the Density of a Gas   184
4.5.7   Partial Pressures   184

Answers to diagnostic test   185

Unit 4.6   General Properties of Solutions   187

4.6.1   Introduction   188
4.6.2   Types of Solution   190
4.6.3   Factors Affecting Solubility   190
         Properties of the solute and solvent   190
         Temperature   190
         Pressure   191

4.6.4   Factors Affecting Rate of Solution   191
        Temperature   191
        Rate of stirring   191
        Particle size   191
4.6.5   Ionic Salts – Solubility Rules   192
4.6.6   Recrystallisation   192
4.6.7   Colligative Properties   193
        Vapour pressure   193
        Raising the boiling point   194
        Lowering the freezing point   194
        Osmotic pressure   194
4.6.8   Colloidal Dispersions   196

Answers to diagnostic test   198

Module 5   Chemical Reactions, Equilibrium and Energy Changes

Unit 5.1   Energy Changes   201

5.1.1   Introduction to Energy Changes   203
5.1.2   Energy Changes in Chemical Reactions   203
5.1.3   Collision Theory   204
        Summary   205
5.1.4   Making and Breaking Bonds   205
5.1.5   Exothermic and Endothermic Reactions   207
        Exothermic reactions   207
        Endothermic reactions   208
5.1.6   The Heat of Reaction   210
5.1.7   Activation Energy   211

Answers to diagnostic test   213
Further questions   214

Unit 5.2   Measuring Rates of Reactions   215

5.2.1   Introduction   217
        Reactions happen at different rates   217
        The units used in the measurement of reaction
        rates   217
5.2.2   Measuring the Rates of Chemical Reactions   218
        The reaction between calcium carbonate and
        hydrochloric acid   218
        Measuring the rate of reaction   219
        Calculating average rates of reaction   220
        The reaction between sodium thiosulfate and
        hydrochloric acid   222
5.2.3   Different Methods used to Monitor the Progress
        of Some Reactions   223

Answers to diagnostic test   224

Unit 5.3   Factors Affecting the Rate of Chemical Reactions   227

5.3.1   Introduction to the Effect of Temperature on Reactions   229
        The reaction between calcium carbonate and
        hydrochloric acid   229

The reaction between sodium thiosulfate and
hydrochloric acid   229
5.3.2   The Effect of Changing the Concentration of Reactants
in Solution   231
5.3.3   The Effects of Pressure Change on the Rate of a
Gaseous Reaction   232
5.3.4   The Effect on the Rate of Reaction of the Surface Area
of Solid Reactants   232
5.3.5   The Effects of Catalysts on the Rate of Reaction   234
What do we know about catalysts?   234
The physical state of catalysts   235
How do catalysts work?   236
'Negative' catalysts   236
Enzymes as catalysts   236
The kinetics of enzyme reactions   238
Enzyme inhibitors   239

Answers to diagnostic test   240

Unit 5.4   Reversible Reactions and Chemical Equilibrium   241

5.4.1   Reversible and Non-reversible Reactions –
Introduction   243
5.4.2   Systems at Equilibrium   244
Physical equilibrium   244
Chemical equilibrium   244
5.4.3   Factors Affecting Systems at Equilibrium   245
Removal of chemicals can affect the reaction   246
Effects of pressure changes on the equilibrium
state in gaseous reactions   246
The effect of temperature on the equilibrium
state   247
The effect of catalysts on the equilibrium state   247
5.4.4   Some Industrial Processes that involve Equilibrium   247
The Haber Process   248
The Contact Process   248
The Ostwald Process   250
5.4.5   The Law of Mass Action   250

Answers to diagnostic test   253

Unit 5.5   Quantitative Aspects of Acid–Base Equilibria   255

5.5.1   Introduction   257
5.5.2   The Dissociation of Water   257
5.5.3   The Measurement of pH, and the pH Scale   258
Calculating values of pH   258
Calculating the pH of alkaline solutions   259
5.5.4   Detecting Acidity and Alkalinity   260
The formation of hydrogen ions, and the acid
dissociation constant   260
Strong and weak acids   261
The use of the pH scale   262
5.5.5   The Effect of Strong Acids and Alkalis on Living
Organisms   262
Buffer solutions   263
How do buffer solutions work?   264

5.5.6   Calculating the pH of a Buffer Solution   265
5.5.7   The Effects of Buffers in the Body   266
        The hydrogencarbonate system   266
        The phosphate buffer system   266
        The amino acids as buffers   267

Answers to diagnostic test   269

5.5.8   Extension – What is the Connection between pH
        and pK_a?   270

Module 6   Compounds of Carbon (1)

Unit 6.1   What's So Special about Carbon Compounds?   275

6.1.1   Introduction – Organic Compounds All Around Us   276
6.1.2   So the Chemistry of Carbon IS Worth Knowing!   278
6.1.3   Making Order of the Millions   278
6.1.4   Carbon in the Periodic Table?   279
6.1.5   Compounds of Carbon and Hydrogen – The Hydrocarbons   280
6.1.6   Summary   281
        Aliphatic and aromatic   281

Answers to diagnostic test   283
Further questions   283

**Unit 6.2   Compounds of Carbon and Hydrogen – Alkanes
            and Alkenes**   285

6.2.1   Alkanes   287
        Naming system   289
        Properties   291
6.2.2   Alkenes   291
6.2.3   Comparison of Properties in the Alkane and
            Alkene Series   293
        Melting and boiling points   293
        What is meant by saturated and unsaturated carbon
            compounds?   295
        Hydrocarbons as fuels . . . combustion   295
        Substitution and addition reactions with halogens   296
        Reaction with hydrogen gas   298
        Reactions with hydrogen chloride or hydrogen
            bromide gas   298
        The cracking of alkanes and the polymerisation
            of alkenes   299

Answers to diagnostic test   303
Further questions   304

6.2.4   Extension – Some Explanations and Mechanisms   306
        The detailed mechanism for the reaction of chlorine
            with methane in bright sunlight   306
        Alkene + HBr mechanism   306

Unit 6.3   Aromatic Organic Chemistry   309

6.3.1   Introduction   310
6.3.2   Methylbenzene (Toluene)   311

6.3.3   Other Compounds of Benzene   312
6.3.4   Substitution Reactions   313
6.3.5   Directing Group Preferences   314

Answers to diagnostic test   316
Further questions   317

Unit 6.4   Isomerism   319

6.4.1   Isomerism   321
6.4.2   Set 1 – Structural Isomerism   321
         Structural isomerism in hydrocarbons   321
         Functional group isomerism (a sub-group of structural
           isomerism)   325
6.4.3   Set 2 – Stereo-isomerism   327
         Optical isomerism   327
         Stereo-isomerism in alkenes – geometric or
           cis–trans isomerism   329

Answers to diagnostic test   331
Further questions   333

Module 7   Compounds of Carbon (2) – Organic Compounds containing Carbon, Hydrogen, Oxygen and Nitrogen

Unit 7.1   The Alcohols, $C_nH_{2n+1}OH$ 337

7.1.1   Introduction   338
7.1.2   Types of Alcohols   339
         Some ethers are isomeric with alcohols   340
7.1.3   Some Reactions of Alcohols   341
         Oxidation   341
         Dehydration   342
         Hydration   343
         Reduction of alcohols   344
         Ester formation   344
         Substitution of an OH group with Cl using PCl_5
           or $SOCl_2$ 345

Answers to diagnostic test   347
Further questions   348

Unit 7.2   Carbonyl Compounds and Carboxylic Acids   349

7.2.1   Carbonyl Compounds – General Formula $C_nH_{2n}O$ 350
7.2.2   Some Properties of Aldehydes and Ketones   350
7.2.3   Some Preparations of Aldehydes and Ketones   352
7.2.4   Reactions of Carbonyl Compounds   352
         Addition reactions   352
         Reduction by addition of hydrogen   355
         Oxidation   355
7.2.5   Carboxylic Acids – General Formula $C_nH_{2n+1}COOH$ 358
         General properties and acidity   358
         Carboxylic acids with metals   358

	Ester formation	359
	Other reactions	359
7.2.6	The Strength of a Carboxylic Acid	359
	Answers to diagnostic test	361
	Further questions	362
	More advanced questions	362
7.2.7	Extension – Aldehydes and Ketones in Sugars	363

Unit 7.3   Amines   365

7.3.1	The Amines	366
	Substitution of the hydrogen atoms of ammonia, NH_3	366
7.3.2	Hydrogen Bonding in Amines	367
7.3.3	Aromatic Amines and Azo Compounds	367
7.3.4	Amino Acids	368
	Condensation to make peptides and proteins	369
7.3.5	Base Strengths of Amines	370
7.3.6	Some Further Reactions	371
	Reactions with acids	371
7.3.7	Summary of Reactions	372
	Answers to diagnostic test	373
	Further questions	374

Unit 7.4   Chemistry in Our Environment   375

7.4.1	General Environmental Issues	376
7.4.2	The Earth and its Atmosphere	377
7.4.3	The Atmosphere	378
	Formation of ozone	378
	Formation of ions	378
	Oxides of nitrogen	378
	Any water or moisture also reacts	378
7.4.4	Photosynthesis	380
7.4.5	Fossil Fuels and Pollution	380
7.4.6	What is the Greenhouse Effect?	381
7.4.7	Acid Rain	382
7.4.8	Acid Rain and Plant Growth	383
7.4.9	Solid Wastes and Other Materials	383
	Some examples of using waste	383
7.4.10	Conclusion	384
	Answers to diagnostic test	385
	Further questions	386

Appendix – Selected Answers to Further Questions   387

Alphabetical List of Elements   390

Greek Symbols   391

Subject Index   392

Aims

The aims, philosophy and content of this text are designed to:

- give an understanding of essential relevant chemical concepts;
- give people (often with limited previous chemical knowledge) sufficient understanding to be able to enter science-related courses involving chemistry in higher education, *e.g.* 'access' or 'foundation' courses, or as a set of course materials for 'subsidiary' courses in chemistry supporting other subjects, *e.g.* biology, environmental science, *etc.*
- present the material in a clear and friendly manner, and to allow the student to self-pace their acquisition of chemical knowledge;
- include relevant applications of chemistry to business, health, sport, industry;
- produce a set of course materials that can be used in a flexible way, from total self-study to guidelines for lecturers.

Format of Materials

The overall course is designed to be used in a flexible way, and the time to complete this text will depend upon your starting knowledge and the time you can devote to study. The time necessary to study each module could be different depending upon the background of the student, but approximately 300 hours of total study time is thought to be sufficient to cover the whole course.

- The material is arranged into seven modules, and the modules are divided into units.
- It will be an effective learning strategy if you divide the units into work sessions lasting about one hour.
- The units start with a list of aims from which you can check your previous or prior knowledge. This is followed by a self-diagnostic test to check the level of understanding. The answers are given later in the text as a check. You should obtain at least 80% on this test to be certain of understanding the material. If the check shows that there is a deficiency in understanding then the work tasks in the unit can be undertaken in detail.
- The understanding of each unit can be checked by repeating the original diagnostic test at the end of the unit. Doing this will reveal whether your understanding has increased. If you still feel there is room for improvement, then repeat the unit at a later date, but sooner rather than later!
- Assessment tasks are included in the text and can be used as 'self-test' items.
- Many worked examples and problems are included in the text where appropriate.
- There is also a set of more advanced questions at the end of some units to check on understanding and provide a challenge, and also to apply the concepts to wider areas. Some of this work might require further reading in other texts. Answers to some of these questions are given in an appendix near the end of the book.
- The work is presented in a variety of styles to cater for the many different learning styles of students. Some tasks involve reading and comprehension, others are thinking-and-doing tasks, and some involve calculations.
- No practical details for doing experiments are included, only the results from some typical experiments. Practical work is left to the discretion of the tutors, if this is used as a complete college-based course. This book is intended as a student study guide.

The material included has been extensively tested with over 2000 students on foundation, access and subsidiary chemistry courses.

Study Guide – how to make the best use of your time

Time and place

Study requires time commitment.

- Plan your week to include time slots that you are going to set aside for the study of this book. Make them high priority periods, which can only be changed in very exceptional circumstances.
- Analyse your degree of alertness or readiness to study, to see which parts of the day are the best for you to spend studying for one-hour, intensive study periods . . . before breakfast . . . lunch-time . . . evening, *etc*. Never study when you know that tiredness will win. This is frustrating and leads to de-motivation.
- Select a suitable place to study, probably with a writing table, but not in a comfortable, sleep-inducing armchair. If a place cannot be found in the home, enquire about the use of a local library; they often have 'reading/study areas'. I know of someone who found that the only place of peace and quiet was in his van, and he drove to a local lay-by and set up his books in the van. It worked for him.
- Study periods should be about one hour in length and it is probably better to have a break between hour-long sessions than to go on for two hours non-stop and drop off to sleep. If you are studying other subjects, then break up the time with interchange of the different subjects.
- Some people who know about physical exercise say that if you are feeling tired you should stand up and jog on the spot for about two minutes; it wakes up the body. A bit difficult in a library! But a brisk 10 minute walk could do the trick.
- Quiet music is said by some to be helpful, but distracting music that might cause you to 'sing along' can be disruptive. Research has shown that stimulating music played for about 10 minutes before a study period, then turned off, can cause your brain to wake up and be ready for study.
- How much studying per week, you ask? The reading and accompanying study will take a considerable time commitment covering a whole year or even two years, but try to cover the work of a unit within a week. Some weeks you might have more time than others, but put aside prime study time to do this work. This book is designed to prepare you in chemistry for entry into a science-related university course, or to give you the basic theory behind any practical chemistry that you use in your work. It is trying to encourage you to become a quality, self-motivated student, not a reluctant or 'forced' student.

Revision and reminders

You should put together a set of your own notes as you go through the modules and units. Your notes together with this book should be studied in a systematic manner; use, say, one hour per week to review the previous chapters. As you do this the material will become much more familiar to you. The start of any study regime can be difficult, but eventually it becomes easier and hopefully enjoyable.

Assignments and essays and self-assessment

It is essential that everyone is able to express themselves in a cogent, logical and clear way. Employers expect it when you write your reports. So it is good to practise these skills yourself, and even to impose an essay on yourself as a means of revision.

Remember that no one is going to 'spoon feed' you on this course; you have to 'feed yourself' if you use the content as a self-study text.

Questions don't have to be set by teachers or lecturers: try writing some questions when you finish a section or unit. Put these to one side and refer back to them when revising the material. Share questions with friends doing the same course. This is like a self-help group. For example, write, say, 500 words or 2–3 pages (no more) on any of the following:

- Discuss how the modern periodic table has been developed.
  or
- What is meant by covalent and ionic bonding.
  or
- Discuss how carbon compounds containing C=C bonds are industrially important.

If you are working in a self-help group, then to mark other people's work you will have to know what you are looking for and how you are going to give the marks. A discussion afterwards of your work and that of others will be invaluable. Make sure the writer sticks to the question asked, NOT what they thought should have been asked, and watch out for deviations from the point of the question.

A tutor with chemical knowledge is often delighted to help someone who shows an interest in the subject, so have your work overseen by such a person.

When attempting to answer questions that require long answers, remember what they told you in other subjects. Plan your essay answers to contain:

- an introduction; followed by
- an account of the main points; and
- end with a short summary.

Increasingly, students are asked to express their views in good English.

The questions in this text are usually of the short-answer type, but you will need to practise the longer answers within a set time limit.

Need for chemists

There is a great need for professional chemists and scientists with sound chemical knowledge and even many non-science-based professions will be enhanced by having people in them with relevant chemical knowledge.

The chemical industry is one of the major income generators for Britain, and its products are one of our major exports. It is because the country needs to educate everyone on chemical matters, and also to prepare people on foundation and access courses for future university science courses involving some chemistry that the Royal Society of Chemistry has initiated this and other books.

Good luck with your studies.

Module 1

Building Blocks of Matter

Unit 1.1
Particles and Properties

Starter check

What you need to know before you start the contents of this unit.

You will need to have an understanding of the general scientific vocabulary, including the meaning of particles, property, solid, liquid, gas, melting, boiling, diffusion.

Aims **By the end of this unit you should understand that:**

- Different substances have different sets of properties.
- There are two sorts of properties, physical and chemical.
- Matter can exist in different states.
- All matter is made up of particles.
- The type, arrangement and movement of the particles affects the properties of matter, including evaporation, condensation, dissolving, diffusion and the effect of pressure on gases.

Diagnostic test

Try this test at the start of the unit. If you score more than 80%, then use this unit as a revision for yourself and scan through the text. If you score less than 80% then work through the text and re-test yourself at the end by using this same test.

The answers are at the end of the unit.

1 List four physical properties of water. **(4)**
2 What is there between the particles of the air we breathe? **(1)**
3 a) In which would you expect the particles to be closer together, a solid or a liquid? **(1)**
  b) In which would you expect the particles to be closer together, a gas or a liquid? **(1)**
  c) In which would you expect the particles to be moving more quickly, a solid or a liquid? **(1)**
  d) In which would you expect the particles to be moving more quickly, a gas or a liquid? **(1)**
4 Complete the following sentences, by using any of the following words. They might be used more than once or not at all.

Repulsion, attractions, moving, vibrating, static, disintegrates, breaks up, remains unchanged, whole, surface, inside, slowest, fastest, upwards, middle, downwards, gas, vapour, fixed, melting, boiling, solid, liquid, quicker, more difficult, easier.

a) When a solid melts, the forces of _attr_ between the particles can no longer hold them together because they are _moving_ too much, so the structure _break up_ and the particles become free to move throughout the _surface_ of the liquid. **(4)**

b) When a liquid evaporates, the _vapour_ moving particles escape from the _surface_ of the liquid. They can escape, because at the surface of the liquid the nett force of attraction on a particle is _slower_ into the liquid. **(3)**

c) When a liquid boils, bubbles are formed in the liquid. These bubbles contain _vapour_. Boiling happens at a _fixe_ temperature called the _Boiling_ point. **(3)**

d) When a solid dissolves in a liquid, particles of the _liquid_ are knocked off by collision with _solid_ particles and they diffuse into and mix with the _liquid_ particles. The hotter the liquid, the _faster_ its particles are moving, so the _melting_ it is for them to knock particles off and dissolve the solid. **(5)**

5 Why do gases diffuse more quickly than liquids? **(2)**

6 For each of the following substances, say whether it would be a solid, a liquid or a gas at room temperature (room temperature is taken as being 20 °C):

Substance A – melting point 25 °C, boiling point 150 °C **(1)**
Substance B – freezing point −25 °C, boiling point 100 °C **(1)**
Substance C – melting point 15 °C, boiling point 110 °C **(1)**
Substance D – freezing point −253 °C, boiling point −150 °C **(1)**
Substance E – melting point 1250 °C, boiling point 1700 °C **(1)**

31 Marks   (80% = 25)

1.1.1 Properties

Chemistry is the study of matter. There are lots of words that are used to mean 'a particular type of matter'. Two of the most common are 'substance' and 'material'. The word 'matter' covers everything the Universe is made of, whether it is solid, liquid or gas. The words 'material' and 'substance' simply mean one particular sort of matter, like water or soil or air or monosodium glutamate.

Look at the paper on which this is printed. It looks white. It is smooth to the touch. Two of the properties of this paper are that it is white and smooth. **Properties** are simply what you can detect or measure about a material or what it does. If you were to put a match to this paper, for example, it would burn. This property is summarised by saying that it is combustible.

Some of the properties of materials are ones that they have **of themselves**, without any other material being involved. Gold, for example, is shiny and melts at 1065 °C. Properties like colour, melting point, boiling point, electrical conductivity, do not depend on any other substance, and are called the **physical properties** of gold.

Some properties of materials describe how they behave with other materials or chemicals. Gold is not attacked by dilute acids, for example, whereas lots of other metals are. Properties like this are called **chemical properties** because a chemical reaction occurs and the material is permanently changed.

Properties are really just descriptions of materials – what they look like, feel like, how they behave, *etc*. Notice that some are negative – 'gold is *not* attacked by dilute acid'. In science, negative information is often just as important as positive.

Each substance has its own unique set of properties.

If you look at a crowd of people: lots of them have dark hair; a few of them may be 2 m tall; lots of them may have a mass of 70 kg; lots of them may have green eyes; lots of them may have eyebrows that meet in the middle, **but** there is probably only one individual in the crowd who has all of these characteristics.

In the same way: lots of substances are colourless; lots of substances are odourless; lots of substances are liquids at room temperature, **but** if you came across a beaker full of a colourless, odourless liquid, particularly in a chemistry lab, you might hesitate to drink it until you'd checked that it wasn't acidic or alkaline and that it boiled at 100 °C. In other words, water, like all other substances, has a **unique** set of properties.

1.1.2 Change of State

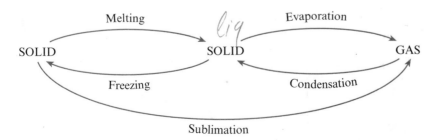

Melting point, boiling point, and freezing point are all properties of substances. The melting point is the temperature at which a substance changes from a solid to a liquid. The freezing point is the temperature at which a substance changes from a liquid to a solid. It is important to understand that these two temperatures are numerically the same – it just depends on whether you are heating it or cooling it, as to what you call them (Figure 1.1.1).

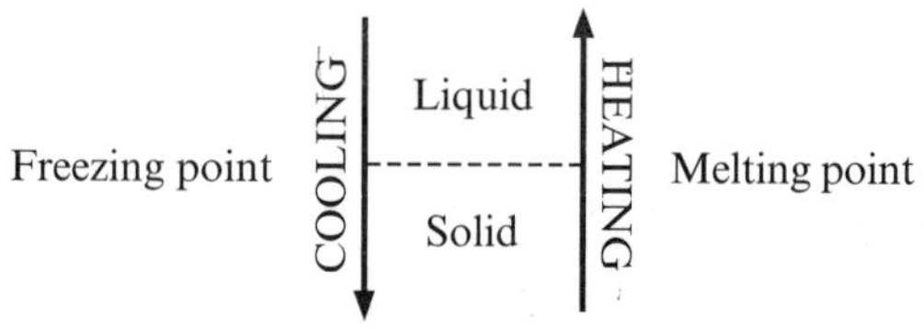

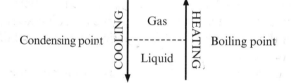

Figure 1.1.1

Change of state between solid and liquid

The same is true for the boiling point and condensing point. Boiling is when the pressure of the vapour trying to escape from the surface of the liquid is just above the pressure of the atmosphere trying to keep it in. So at the boiling point a lot of the vapour escapes and bubbles of vapour can be seen; at temperatures well below this, only a little of it evaporates as only a few particles have enough energy to break through the surface (Figure 1.1.2).

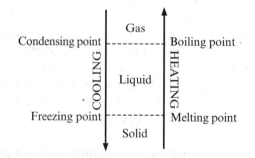

Figure 1.1.2

Change of state between liquid and gas

Putting these together, you get a diagram shown in Figure 1.1.3:

Figure 1.1.3

Changes of state between solid, liquid and gas

Suppose you know the melting and boiling points of a substance, you can decide whether it will be a solid, a liquid or a gas at any temperature by using the diagram. For example, if you had a substance that has a melting point of 10 °C and a boiling point of 98 °C, what state will it be in at 56 °C?

As you can see from Figure 1.1.4, 56 °C is between the melting and the boiling point, so the substance will be a liquid at that temperature because the temperature of its boiling point has not yet been reached. But even at temperatures below its boiling point, some of the liquid can slowly evaporate away.

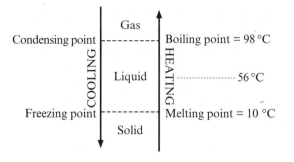

Figure 1.1.4
Example substance is a liquid at
56 °C

1.1.3 Particles

It is sometimes difficult to believe that everything is made up of particles, because they are so small that you can't see them. In several places around the country, *e.g.* Wiltshire and Oxfordshire, there are figures carved into the chalk hillside.

The carvings are made up of individual little bits of chalk, but when we are a long way from it they are so small that we cannot see them – all we see is the smooth horse-shaped figure. In the same way the edge of this paper looks perfectly smooth, but if you look at it with a powerful magnifying glass or microscope, you will see that it is actually quite ragged. It's just that the indentations are so small they cannot be seen with the naked eye.

Even though it hasn't been possible to see such particles until fairly recently, scientists have believed for a long time that all matter is made up of particles. Understanding how these particles behave explains some of the properties of matter.

Before we look at these explanations, it is important to establish one thing. When we say that all matter is made up of particles, it raises the question 'what is there in between the particles?' In a recent survey conducted in the fictitious town of West Framlingham, we asked people what they thought there was between the particles of air. It was noted that when asked the question, one hundred percent of those polled said nothing (in fact their mouths fell open and a blank stare appeared in their eyes). And they were absolutely right to say 'nothing' because there is 'nothing' at all in between the particles. (Actually, one of the respondents was pressed to speak and said 'there was even smaller bits of air' and another said, 'glue keeps them together'.) We explained that '**all** matter' includes 'air and glue' and that these must also be made up of particles. We asked again what was between their particles, and they soon got the point and now said NOTHING. Right again.

So what properties of materials does the existence of particles explain?

STATE AND CHANGE OF STATE

What are the properties that distinguish between solids, liquids and gases? Firstly, solids have a fixed shape, whereas liquids and gases change shape spontaneously – they 'flow' and so must be kept in a container.

Secondly, solids and liquids have a fixed volume, whereas gases expand spontaneously to take up all the space that is available to them.

Thirdly, solids are almost impossible to compress or squeeze, liquids are very difficult, even in a sealed syringe, but gases are easy to compress and squash (think of a bicycle pump where gas is squashed into a tyre).

How can we explain these properties in terms of particles?

In a solid, the particles are fixed in place in some kind of order. They **can** move, but only by vibrating backwards and forwards around fixed positions. You could imagine it as being like a scrum in a game of rugby where the players are locked together, tugging and heaving but not letting go of each other (Figure 1.1.5).

Figure 1.1.5
Rugby scrum

Of course, in a solid the particles aren't held together by putting their arms round each other. What holds them in place are **forces of attraction** between the particles.

Still thinking about the scrum, think about what happens when the ball is thrown into it. The players start to move more vigorously (not to mention violently) as they jostle for position and possession. As the movement increases in violence, players start to lose their grip on each other. They become free to move independently. It is the same with the particles in a solid. If you put energy into the solid, the particles start to vibrate more vigorously.

The forces of attraction cannot hold them together any longer and the particles become free to move independently. They move in a straight line until they collide with something and bounce off. The solid melts (Figure 1.1.6).

The particles in the liquid aren't much further apart than they were in the solid, so you can compress liquids a bit (a little like a loose maul in rugby), but not much because the particles are quite close together already.

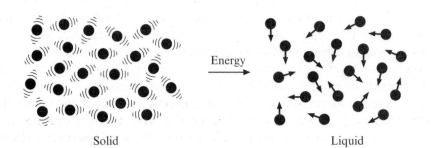

Figure 1.1.6
Melting of a solid

Solid   Energy   Liquid

To go back to rugby: once the ball is free from the scrum, the players who were in the scrum move more widely apart and play speeds up. This is just like what happens when you heat a liquid. Its particles start to move so quickly that the forces of attraction are completely overcome, the particles are no longer restrained at all; they move as far apart as they can and the liquid evaporates or boils, and the gas spreads out to occupy all the space available in the room or wider afield.

Suppose you had two teams playing, one made up of really massive players and one of smaller players. Which ones would you have to apply more force to, to get them moving about in the scrum? Obviously the answer is the more massive ones. It is the same with substances. Other things being equal, substances with more massive particles will have higher melting and boiling points, because you need to put in more heat energy to get their particles moving fast enough to break up the solid structure or escape from the liquid.

Suppose you had two teams playing whose players were equally massive, but one team had tremendous arm strength and could lock together really well. Which ones would you have to apply more force to now to break them apart from the scrum? Obviously it's the ones with the strongest grip on each other. It is the same with substances. Other things being equal, the particles that have the strongest forces of attraction between them will have the highest melting and boiling points, because you need to put in more heat energy to get their particles moving fast enough to break up the solid structure or escape from the liquid.

1.1.4 Evaporation

There are two ways that liquids can change to gases: boiling, and evaporation. What are the differences between them?

When a liquid boils, you can see bubbles in the body of the liquid. These are **not** full of air. They are full of the vapour of the substance that is boiling. Boiling happens in the **body of, or inside** the liquid, but with much turbulence and sufficient energy to push out a lot of particles through the surface into the vapour. When a liquid evaporates, you don't see anything in the inside of the liquid. Evaporation happens at the **surface** of the liquid. Even solids evaporate; for example if you put a solid block of 'air freshener' in a room the smell can be noticed from some distance away, eventually. That is because some of the particles, usually the loosely-held and smaller ones, are evaporating from the block, pass into the air and eventually reach your nose. It shows that some solids can evaporate. Some solids do not evaporate, because their particles are held strongly together and might anyway be too large to evaporate. The change from a solid *directly* to a gas, or from a gas to a solid, without going through the liquid state, is called **sublimation**.

Liquids can evaporate slowly or they can be made to boil. The boiling point of a liquid is the temperature at which boiling occurs. If water is heated, its temperature will start to go up until it reaches 100 °C, and it will stay at that temperature while the liquid boils. Boiling happens at a **fixed temperature** called the **boiling point**. At this temperature there are loads of water particles breaking through the surface of the water as they overcome the pressure of the air trying to keep them in. At the boiling point the **vapour pressure** of the water escaping from the liquid equals the atmospheric pressure. Evaporation, on the other hand, happens at any temperature. Even water left outside at only 1 °C would evaporate slowly

DON'T confuse EVAPORATION and BOILING! Look for large bubbles in the body of the liquid . . .

Unit 1.2
Inside the Atom

Starter check

What you need to know before you start the contents of this unit.

A general scientific vocabulary and the contents of Unit 1.1.

Aims **By the end of this unit you should understand that:**

- Elements and compounds are different.
- Elements are made up of atoms.
- Atoms have a structure.

Diagnostic test

Try this test at the start of the unit. If you score more than 80%, then use this unit as a revision for yourself and scan through the text. If you score less than 80% then work through the text and re-test yourself at the end by using this same test.

The answers are at the end of the unit.

1 Which of the following **must** be compounds:

   a) Substance A, a green solid that goes black on heating and gives off a colourless gas.
   b) Substance B, a yellow solid that melts on heating to form an amber coloured liquid that turns back to a yellow solid on cooling.
   c) Substance C, a white solid that turns yellow on heating and goes back to white on cooling.
   d) Substance D, a white solid that dissolves in water with the production of a lot of heat.
   e) Substance E, a white solid that stays white on heating but gives off a colourless gas. (5)

2 What does the atomic number of an element tell you about the structure of the atom? (2)

3 If the atomic number of an element is 13, how many electrons must there be in an electrically neutral atom of the element? (1)

4 Draw diagrams to show the electronic configurations of the following elements:

   a) carbon (atomic number 6); (3)
   b) magnesium (atomic number 12); (4)
   c) chlorine (atomic number 17); (4)
   d) calcium (atomic number 20). (5)

24 Marks   (80% = 19)

1.2.1 Elements

Most of the substances that we come across every day are not **pure**. In everyday language, 'pure' simply means 'has nothing in it that should not be there'; in science, it means 'no material other than itself'.

Most substances are mixtures. Even the water that comes out of your taps isn't a chemically pure substance. It has got chlorine dissolved in it to kill harmful micro-organisms. In hard water areas it has chemicals from the rocks dissolved in it. It is a mixture. To get pure water you have to distil it: this drives only steam from the solution and condenses it back again into the single substance – the pure water (Figure 1.2.1).

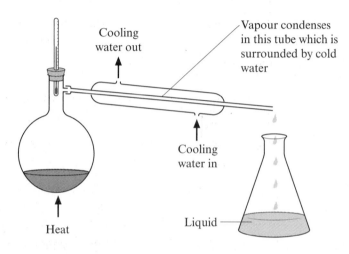

Cooling water out

Vapour condenses in this tube which is surrounded by cold water

Cooling water in

Heat

Liquid

Figure 1.2.1
Simple distillation apparatus

Air is not scientifically pure as it is a mixture of gases. These can be separated by first of all condensing it to a liquid (by cooling it to about −200 °C or 73 K), and then allowing the mixture to boil (you don't need to heat it, just let it warm up at room temperature). The different gases have different boiling points, which enable them to be separated by boiling – this is called fractional distillation. These separated gases have many applications, for example, as in Figure 1.2.2.

The Ancient Greeks used to think that there were only four essential substances, and that everything else was made up from them. They called them the four elements: earth, air, fire and water. Nowadays we know that two of these (air and earth) are not themselves pure substances, but

Figure 1.2.2
Charging BOC's liquid oxygen into the Lox pots used to store liquid oxygen on an RAF Tornado at RAF Leeming, North Yorkshire, UK. The oxygen is used in the Tornado's breathing oxygen system.

mixtures. We know that fire isn't a substance at all. Water is a single substance. But we still retain the word 'element' to mean a single, uncombined substance that cannot be broken down into anything more simple.

Even when you get a pure substance, like distilled water, it isn't necessarily a **simple** substance made up of one element. If you pass electricity through water, for example, the water splits up into hydrogen gas and oxygen gas. If you melt pure salt (800 °C or 1073 K) and pass electricity through it, it splits up into sodium metal and chlorine gas.

On the other hand, it doesn't matter what you do to hydrogen, oxygen, sodium and chlorine – they never split up further. They are what we nowadays call **elements**; they are single substances that cannot be broken down into anything more simple.

1.2.2 Compounds

There are roughly 90 naturally-occurring elements on Earth and everything else is made up of them. Elements are substances that cannot be split up into anything simpler. Substances that can be split up are not simple – they are compound substances. That is usually shortened to just **compound**. So water is a compound of hydrogen and oxygen. Salt is a compound of sodium and chlorine. Compounds always have exactly the same composition of the elements in them no matter how these compounds were made. They have a constant composition. (See also later when this will be quantitatively explained in Module 4, Section 4.2.1.)

Whenever you split up water, and wherever the water comes from, you **always** get the same proportions of hydrogen and oxygen.

It doesn't matter where the water comes from, you *always* get twice as much hydrogen as oxygen, by volume, when you split it up (Figure 1.2.3). If you measure the mass of hydrogen and oxygen, water always contains 11.1% hydrogen and 89.9% oxygen. How can this be explained in terms of particles?

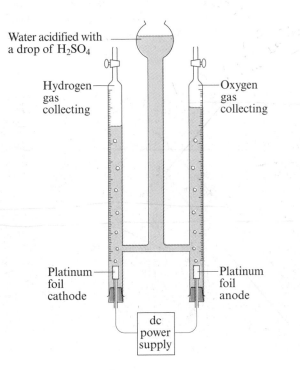

Water acidified with
a drop of H_2SO_4

Hydrogen gas collecting

Oxygen gas collecting

Platinum foil cathode

Platinum foil anode

dc power supply

Figure 1.2.3

Electrolysis apparatus

1.2.3 Atoms

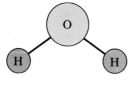

Figure 1.2.4
Water molecule

Elements are the simplest substances that can exist free in nature, *e.g.* a piece of copper metal. They are made up of simple particles, called **atoms**. Hydrogen is made up of hydrogen atoms and oxygen is made up of oxygen atoms. Water is made up of hydrogen atoms and oxygen atoms that are firmly joined together to make water compound particles. One particle contains two hydrogen atoms joined to one oxygen atom to make a water **molecule** (Figure 1.2.4). A molecule is a particle made up of two or more atoms which are firmly held together. You can get molecules of two of the same atoms held together, *e.g.* oxygen gas is O joined to another O, *i.e.* O_2, or you can get two (or more) different atoms joined to form molecules; *e.g.* water is H_2O, where two atoms of hydrogen combine with one atom of oxygen.

All water molecules are like this, so you always get the same proportions of oxygen and hydrogen when you split water up. All other compounds are similar; they are made up of the atoms of two or more elements bonded together. There are millions of different compounds, but they are *all* made up of combinations of some of the 90 or so elements.

For a long time this picture was the one that scientists had of the types of particles making up all matter. The basic building block was considered to be the atom; there are only about 90 different elements, so there are only about 90 different sorts of atom. Atoms were thought to be indivisible. There was no explanation for why two atoms of hydrogen always joined with one atom of oxygen. The scientists asked, 'Why not three or one? Are there any rules to help know how elements join together?' After we have considered the next few sections you will be able to answer these questions.

1.2.4 Smaller Particles

The first hint of the answer came in 1897, when J. J. Thompson showed that atoms were not indivisible, they could be broken open; they are made up of even smaller particles. The first ones to be carefully thought about, by Stoney in 1874, were called **electrons**. These are even smaller than atoms. The lightest atoms (hydrogen atoms) are almost two thousand times as massive as electrons.

Another difference between atoms and electrons is that electrons are negatively charged, whereas atoms are always electrically neutral. But if atoms contain electrons, and electrons are negative, how can atoms be neutral?

The answer is that there are positive particles in an atom as well. These positive particles are called **protons**. The positive charge on a proton is just big enough to cancel out the negative charge on an electron. The simplest atoms (hydrogen) are made up of just one proton and one electron. Since electrons are only about one two-thousandth of the mass of a hydrogen atom, and hydrogen atoms only contain one proton and one electron, it follows that a proton is about two thousand times as massive as an electron.

How are the protons and electrons arranged inside the atom? How are they held together? Continue reading!!!!

1.2.5 Electronic Configuration

The way the smaller particles are arranged inside an atom is called the **configuration** of the atom. The very centre of the atom is called its **nucleus**. The protons are found in the nucleus, so the nucleus is positively charged. In the simplest picture of the atom, the electrons orbit the nucleus, just like the planets orbit the Sun. The planets stay in their orbits because they are attracted towards the Sun by the force of gravity. In the same way the negative electrons stay in their orbits, moving at extremely fast speeds but held from escaping completely because they are attracted towards the positive nucleus (Figure 1.2.5).

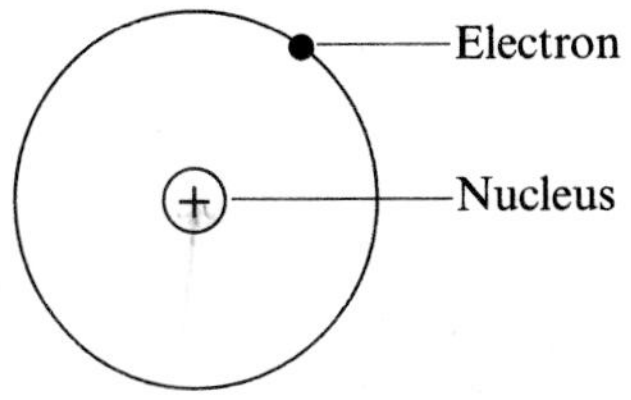

Figure 1.2.5
Electrons orbit the nucleus

Between the nucleus and the electrons there is absolutely nothing – just empty space; quite a lot of it. If you imagine the nucleus as being the same size as the centre spot at Wembley Stadium, the nearest electron would be the size of a peanut in the stands.

Each element has a different number of protons in its nucleus. The number of protons is called the **atomic number** of the element. Sodium, for example, has atomic number 11, which means each sodium atom has 11 protons in its nucleus. If it has 11 protons, of course, it must have 11 electrons to make it neutral.

In the Solar System, each planet follows a unique orbit around the Sun. It's not quite the same inside an atom. Each orbit can hold more than one electron. The first orbit can hold two electrons.

The simplest atoms are hydrogen atoms. The atomic number of hydrogen is 1; each atom has one proton and one electron.

The next simplest atoms are helium (He) atoms with atomic number 2; two protons and two electrons (Figure 1.2.6). This first shell of electrons is small and is completely filled by only two electrons.

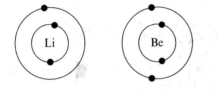

Figure 1.2.6
Hydrogen and helium

Then come lithium and beryllium, with atomic numbers 3 and 4, respectively (Figure 1.2.7):

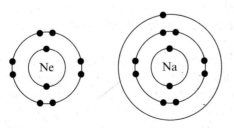

Figure 1.2.7
Lithium and beryllium

The second orbit, being a little larger, can hold a maximum of eight electrons. So the element of atomic number 10 (neon) will have a completely full outer electron shell or orbit containing eight electrons; sodium (atomic number 11) has one electron in the third shell (Figure 1.2.8).

Figure 1.2.8
Neon and sodium

Similarly, once there are eight electrons in the third shell, this also becomes completely full and the next electrons go into the fourth shell. Thus, for the electronic configurations of argon and potassium (atomic numbers 18 and 19, respectively) each electron shell can be shown in the following way, 2:8:8 for argon and 2:8:8:1 for potassium (Figure 1.2.9).

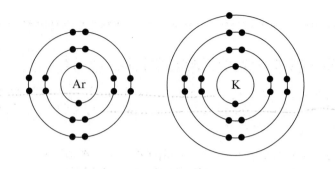

Figure 1.2.9
Argon and potassium

This electronic configuration helps us to explain the chemical properties of matter in much more detail, as will be seen from Unit 1.3.

1.2.6 Further Reading

Read about the work of Rutherford (he was the first to show that the atom had a nucleus and external electrons), Millikan (he measured the charge on the electron), and Bohr (he was the first person to put forward the idea of electrons in orbits).

Answers to diagnostic test

If you score less than 80%, then work through the text and re-test yourself at the end using this same test. If you still get a low score then re-work the unit at a later date.

1   Substances A and E are the only ones that **must** be compounds, because they clearly are composed of more than one element because something was 'given off' and something different was left. The others changed back to the starting materials and nothing new was given off or made.
    5 marks if you put just these two – knock off one mark for every extra one you put down as being a compound, or for each one of A and E you didn't put down.   **(5)**

2   The number of protons in the nucleus of its atoms, and the number of electrons in orbits around the nucleus.   **(2)**

3   13   **(1)**

4   a) Your diagram should show 2 orbits (1 mark), with 2 electrons in the one nearest the nucleus (1 mark) and 4 electrons in the one furthest away from the nucleus (1 mark), or 2:4.   **(3)**

    b) Your diagram should show 3 orbits (1 mark), with 2 electrons in the one nearest the nucleus (1 mark), 8 electrons in the middle orbit (1 mark) and 2 electrons in the one furthest away from the nucleus (1 mark), or 2:8:2.   **(4)**

    c) Your diagram should show 3 orbits (1 mark), with 2 electrons in the one nearest the nucleus (1 mark), 8 electrons in the middle orbit (1 mark) and 7 electrons in the one furthest away from the nucleus (1 mark), or 2:8:7.   **(4)**

    d) Your diagram should show 4 orbits (1 mark), with 2 electrons in the one nearest the nucleus (1 mark), 8 electrons in the second orbit (1 mark), 8 electrons in the third orbit (1 mark) and 2 electrons in the one furthest away from the nucleus (1 mark), or 2:8:8:2.   **(5)**

24 Marks   (80% = 19)

FURTHER QUESTIONS (NO ANSWERS GIVEN)

1   In the text it said that there are about 90 naturally-occurring elements, but when you look at a list of elements there are often 106 or more listed. How can you explain that?

2   The early Greeks, some 2000 years ago, had some ideas that were quite amazing for their time. Look up the philosopher Democritus and his ideas of the atom.

Unit 1.3
Atoms, Elements and Order

Starter check

What you need to know before you start the contents of this unit.

You should already know about what is meant by particles, and that materials are made up of atoms and molecules.

Aims **By the end of this unit you should understand that:**

- The mass of an atom is almost all in the nucleus.
- The nucleus of atoms other than hydrogen contains two types of particles.
- Not all atoms of one element are the same in all respects.
- Elements are represented by symbols.
- The elements are classified in the Periodic Table.

Diagnostic test

Try this test at the start of the unit. If you score more than 80%, then use this unit as a revision for yourself and scan through the text. If you score less than 80% then work through the text and re-test yourself at the end by using this same test.

The answers are at the end of the unit.

1 State how many of each type of particle there are in the following elements:

   a) carbon, atomic mass 12, atomic number 6;
   b) fluorine, atomic mass 19, atomic number 9;
   c) magnesium, atomic mass 24, atomic number 12;
   d) phosphorus, atomic mass 31, atomic number 15. **(6)**

2 What will be the electronic configuration of each of the elements in Question 1 above? **(8)**

3 What is meant by the term 'isotope'? **(2)**

4 Whose Periodic Classification of the elements formed the basis of our modern day periodic table? **(1)**

5 On which side of the periodic table are the metals found? **(1)**

6 What group names are given to the Group 1, Group 2, Group 7 and Group 0 elements? **(4)**

7 What was the first periodic table based on, and what is the modern one based on? **(2)**

24 Marks (80% = 19)

1.3.1 Three Elementary Particles

Name of particle	Symbol	Relative mass	Charge
proton	p	1	+1
neutron	n	1	0
electron	e	very small	−1

Unit 1.2 showed that atoms contain electrons and protons. Protons are about two thousand times more massive than electrons. Protons are in the nucleus, and electrons in the 'orbits' around the nucleus.

The simplest atoms (hydrogen) have only one proton. The next simplest (helium) have two. You might expect, then, that helium atoms would be twice as massive as hydrogen atoms. When you actually compare their masses, you find that helium atoms are four times as massive as hydrogen atoms. Why is this? There cannot be more electrically-charged particles in the helium because this would mean that it is no longer helium.

There is another particle that goes to make up atoms. It is called the **neutron**. As its name suggests, it is electrically neutral. It is found in the nucleus, and it has the same mass as a proton. This particle was suspected to exist for a number of years but it was not confirmed until James Chadwick did some experiments in 1932 and discovered it. This made all the difference to working out the modern idea of the structure of the nucleus of atoms.

The atomic number of hydrogen is one; each atom has one proton in its nucleus. The atomic number of helium is two; each atom has two protons in its nucleus. Helium atoms are four times as massive as hydrogen atoms because they have two neutrons in the nucleus as well, and neutrons have nearly the same mass as a proton.

Just as the atomic number of an element tells you how many protons there are in the nucleus, so the **mass number** tells you how many particles in the nucleus altogether (the mass of the electron is negligible as compared with the protons and neutrons).

The atomic number of hydrogen is 1, and its mass number is 1, because there is just one proton in the nucleus (Figure 1.3.1).

The atomic number of helium is 2, and its mass number is 4, because there are two protons and two neutrons in the nucleus.

You will often see helium (Figure 1.3.2) written as $^{4}_{2}He$. The atomic number is at the bottom and the mass number is at the top.

⊕

Figure 1.3.1   Hydrogen nucleus

Figure 1.3.2   Helium nucleus

1.3.2 Symbols

'He' is the symbol for helium. All elements are represented by symbols. Mostly the symbol is a two letter abbreviation; very often they are the first two letters of the element's name.

Sometimes, particularly for the elements that have been known for a long time, the symbol is just the first letter of its name. Sometimes this rule won't work at all because another element has already been given that abbreviation, and the symbol has to be based on the Latin name for the element.

For example, carbon is 'C'. Cobalt is 'Co'. The first two letters of 'copper' are the same as for 'cobalt', so the symbol for copper is 'Cu', which is based on its Latin name 'cuprum'.

It is important to remember that the first letter of the symbol is always a capital and the second letter is always small. If you're trying to write the symbol for cobalt, and you write 'CO', you aren't writing 'cobalt' at all. 'C' is carbon and 'O' is oxygen, so CO is 'carbon monoxide', **not** 'cobalt'. Although there are rules for the symbols, there is no easy way of working them out. But don't worry, the more you use them, the more familiar with them you'll become. You can see all the symbols for the elements in the periodic table near the front of the book.

1.3.3 Atomic and Mass Numbers

As noted, the symbol for an element is simply either one or two letters. Sometimes it is necessary to show the atomic and mass numbers of the elements as well. When you see the symbol for an element written out with the atomic and mass numbers, like helium shown earlier, you can work out how many of each type of particle there are in a neutral atom of the element. Sodium, for example, is shown in Figure 1.3.3.

The atomic number is 11, so there must be 11 protons. If there are 11 protons, there must be 11 electrons to make the atom neutral. The mass number is 23, so there are 23 particles in the nucleus altogether. Eleven of them are protons, so the other 12 must be neutrons. You can go further still. If there are 11 electrons, two of them must be in the first orbit, eight in the second and one in the third, or 2:8:1.

Similarly, chlorine (see Figure 1.3.4) can be written as $^{35}_{17}Cl$.

The atomic number is 17, so there must be 17 protons. If there are 17 protons, there must be 17 electrons to make the atom neutral. The mass number is 35, so there are 35 particles in the nucleus altogether. Seventeen of them are protons, so the other 18 must be neutrons. You can go further still. If there are 17 electrons, two of them must be in the first orbit, eight in the second and seven in the third, or 2:8:7.

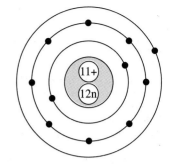

Figure 1.3.3
Sodium, $^{23}_{11}Na$

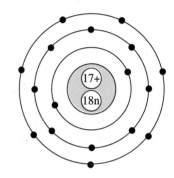

Figure 1.3.4
Chlorine, $^{35}_{17}Cl$

1.3.4 Isotopes and Relative Atomic Mass

Not all atoms of chlorine have got 18 neutrons in their nucleus. Some chlorine atoms have got 20 neutrons. **All** neutral chlorine atoms must have 17 protons and 17 electrons. Atoms that have the same atomic number (number of protons) but different mass numbers (particles in the nucleus) are called **isotopes**. So chlorine has two isotopes (Figure 1.3.5):

$$^{35}_{17}Cl \qquad\qquad ^{37}_{17}Cl$$

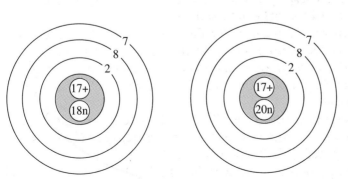

Figure 1.3.5
Isotopes of chlorine

It doesn't matter where a sample of chlorine comes from, it always contains these two isotopes in the same proportions; in fact there are always three times as many ^{35}Cl atoms as there are ^{37}Cl atoms. When you do any mass measurements on chlorine, then, the answer you get is always

35.5 – the weighted average of the two. We say that the **relative atomic mass** (RAM) of chlorine is 35.5. It is the same for other elements. Because they have isotopes, their RAMs are not whole numbers.

The mass of an atom of chlorine, or any other element, is measured relative to something, just as we measure length using a standard measured metre rule. The standard has to be agreed worldwide and has to be constant, so the isotope of carbon with a mass number of 12 was chosen to be the standard, and everything is measured relative to $\frac{1}{12}$ of the mass of the carbon-12 isotope.

The modern periodic table is a classification of the elements according to atomic number rather than RAM.

Near the front of this book you'll find one of the most important classification systems in science: the Periodic Table of the elements, which includes the atomic numbers and mass numbers.

1.3.5 Periodicity

When the idea of relative atomic mass was first put forward, some chemists started to wonder whether there was a connection between the RAM of an element and its properties. One of these chemists, Dobereiner, in 1829, noticed that there were groups of three elements (triads) which had very similar chemical properties and in which the RAM of the middle element was almost exactly the average of the other two elements in the triad (these groups became known as Dobereiner's Triads). One triad was the three elements lithium, sodium and potassium; all are soft metals, are highly reactive and have to be stored under oil. Their RAMs are: lithium 7, sodium 23, potassium 39. As you can see, sodium's RAM is the average of 7 and 39.

Similarly for chlorine (35.5), bromine (80) and iodine (127), all of which are coloured, poisonous and dissolve in water to give acidic solutions.

It was noticed that some physical properties of the elements such as melting point rise and fall in a regular way as RAM increases. Such a variation is called a **periodic variation**, because it varies periodically. The first chemist to try plotting a graph of property against RAM, about a hundred and forty years ago, Lothar Meyer, plotted the 'atomic volume' of elements, *i.e.* atomic mass/density, against RAM, and he noticed a periodic pattern in the graph which he linked to the properties of the elements.

You might like to plot the graph using a more modern set of data. Plot the atomic radius of the atom against its RAM. The values are given in Table 1.3.1. Do you see any patterns?

Table 1.3.1 *Table of atomic radii* (pm)/*RAM*

^{1}H							^{4}He
32							50
^{7}Li	^{9}Be	^{11}B	^{12}C	^{14}N	^{16}O	^{19}F	^{20}Ne
152	112	98	91	92	73	72	70
^{23}Na	^{24}Mg	^{27}Al	^{28}Si	^{31}P	^{32}S	$^{35.5}$Cl	^{49}Ar
186	160	143	132	128	127	99	98
^{39}K	^{40}Ca						
227	197						

RAM

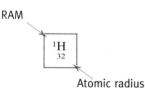

Atomic radius

The real problem for these early workers was the business of undiscovered elements, particularly for Newlands. Newlands arranged the elements in order of increasing RAM, as shown:

H Li Be B C N O F Na Mg Al Si *etc.*

Newlands noticed that in this arrangement, elements with similar properties occurred every eight places (start counting with '1' at Li and when you get to '8' you'll be on sodium, which is very similar to lithium). Newlands called this the 'Law of Octaves'. The only trouble was that further along the list, it stopped working. Newlands didn't realise it, but this was because some elements had not yet been discovered, so it threw his system out of order. It is funny how difficult it is for anyone at any time not to assume that almost everything there is to know has been discovered. Looking back now, it's easy to see why the order broke down, but it took a genius to realise it. The name of the genius was Dmitri Mendeleev, a Russian chemist.

Mendeleev published his first ideas of a system for classifying the elements in March 1869, and he did more or less what Newlands had done, with two exceptions. Firstly, he arranged the elements in a table (still strictly in order of increasing RAM), so that elements with similar chemical properties fell under one another:

H
Li Be B   C   N   O   F
Na Mg Al Si P   S   Cl
K   Ca Ga . . .

After Gallium, Ga, the next **known** element in order of RAM was arsenic (As). But arsenic is nothing like carbon and silicon in its properties. It is, on the other hand, a lot like nitrogen and phosphorus. So the second difference between Mendeleev and Newlands is that Mendeleev put arsenic under nitrogen and phosphorus and left a gap under carbon and silicon.

Now for the real genius part: he said that an element would eventually be discovered which would have a RAM and properties that would mean that it fitted into the gap he had left. He even predicted its properties – the properties of an element that had never been seen by anyone. Only a few years later an element was discovered and given the name 'germanium' (Ge). When its RAM was measured, it fitted exactly between gallium and arsenic. When its other properties were studied, it was found that they fitted Mendeleev's predictions with amazing accuracy (he had predicted its density to within three decimal places, for example). This and other predictions were sufficient to convince other scientists of the validity of the periodic classification of the elements, but there were still a few anomalies to be sorted out, which were not fully appreciated until 1932 when the neutron was discovered.

In the periodic table, families of elements with very similar properties fall under one another and are called '**Groups**'. The groups are numbered, starting with Group 1 on the left-hand side. If you look at the diagram on page 30 you'll notice that the group on the extreme right hand-side is given the number zero. That is because that particular family of elements do not seem to have any pronounced chemical reactions. For this reason they were once called the 'Inert Gases' but are now called the 'Noble Gases'. Some other groups also have names. The Group 1 elements were called the Alkali Metals; Group 2 are called the Alkaline Earth Metals, and Group 7 are called the Halogens. These older names have lasted over the years.

Rows of elements are called '**Periods**'. Notice how, as you go across a period, the elements on the left are metals and will react with water

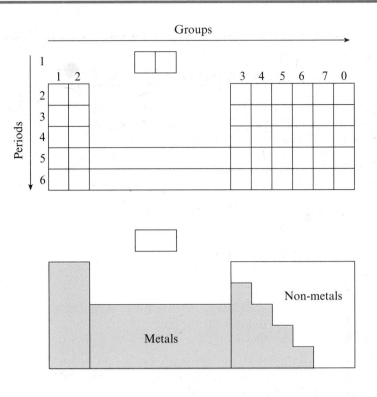

Figure 1.3.6
Storing sodium metal

and/or acids to give hydrogen gas, whereas the elements on the right are non-metals and react with water to form acids or with alkalis to form salts. Between them there are some elements called 'metalloids'. They have some metallic and some non-metallic properties.

If you look at a group in the middle of the table, you'll see that the elements at the top of the group are non-metals, and the ones at the bottom are metals, so somewhere in the middle of the periodic table are the elements that seem to have a dual 'personality'. It is these elements (germanium, silicon and arsenic) that have become so useful in making the modern materials for electronic integrated circuits.

The Group 1 metals (or alkali metals) for example, are all very soft metals (they can be cut with a knife); they all react vigorously with cold water to produce hydrogen and form an alkaline solution of the metallic hydroxide (hence the name alkali metal); they all have to be stored under oil to prevent them reacting spontaneously with the oxygen and water vapour in the air (Figure 1.3.6), *e.g.* for sodium:

$$4Na(s) + O_2(g) \rightarrow 2Na_2O(s).$$

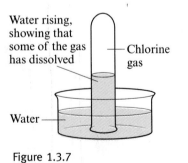

Figure 1.3.7
Chlorine

Group 7, or the halogens, are the most non-metallic elements, with fluorine being the most non-metallic element of all – so much so that it reacts with almost anything and is a difficult element to work with. Its compounds on the other hand are some of the most stable and difficult to decompose. But consider chlorine (Figure 1.3.7): it is a yellow poisonous gas which dissolves in water to form an acid solution,

$$Cl_2(g) + H_2O(l) \rightarrow HCl(aq) + HClO(aq)$$

and it readily reacts with alkalis to form salts. With sodium hydroxide,

$$Cl_2(g) + 2NaOH(aq) \rightarrow NaCl(aq) + NaClO(aq) + H_2O(l)$$

So the periodic table contains a wealth of chemical information if you know how to use it.

Answers to diagnostic test

If you score less than 80%, then work through the text and re-test yourself at the end using this same test. If you still get a low score then re-work the unit at a later date.

1   Protons Electrons Neutrons
  a) 6   6   6
  b) 9   9   10
  c) 12   12   12
  d) 15   15   16
  Half a mark for each correct.   **(6)**

2 a) 2:4
  b) 2:7
  c) 2:8:2
  d) 2:8:5
  2 marks for each correct.   **(8)**

3 Atoms having the same number of protons but different numbers of neutrons, or atoms having the same atomic number but different mass numbers.   **(2)**

4 Dmitri Mendeleev   **(1)**

5 The left.   **(1)**
  Bonus if you say also in the transition series or 'd'-block

6 Group I   The Alkali Metals
  Group II   The Alkaline Earth Metals
  Group VII The Halogens
  Group 0   The Noble or Inert Gases   **(4)**

7 The first was based on relative atomic mass. The modern is based on atomic number.   **(2)**

24 Marks   (80% = 19)

FURTHER QUESTIONS (NO ANSWERS GIVEN)

1 Give the symbols of the following elements:
  Iron   Lead   Potassium
  Krypton   Iodine Sulfur
2 Which of these elements would be the most reactive towards oxygen in the air?
3 Which element is the most unreactive?
4 Which elements are metals and which are non-metals?
5 Which element has 26 protons in its nucleus? How many neutrons will it have if the atomic mass is 56?
6 What do you call a substance that also has 26 protons but has an atomic mass of 55? (See Question 5.)
7 How many electrons will potassium have in the outermost electron shell? In which group of the periodic table will potassium be found?
8 How would you expect potassium and iodine to react together? What will be the compound formed? How might each gain a stable outer electronic structure?

9 What would be the compound that lead and sulfur would form if reacted together?

10 What would be the formula of the compound formed if potassium and sulfur reacted together?

11 The history of the development of a system for assigning symbols for the elements is fascinating, and you might like to look at the work of Dalton.

Unit 1.4
The Nucleus of the Atom

Starter check

What you need to know before you start the contents of this unit.

You should have an understanding of the structure of the atom as outlined in Units 1.2 and 1.3.

Aims **By the end of this unit you should understand that:**

- The nucleus contains protons and neutrons.
- The ratio of protons to neutrons seems to determine the stability of the nucleus. Unstable nuclei can split up.
- The splitting of nuclei causes emission of radiation.

Diagnostic test

Try this test at the start of the unit. If you score more than 80% then use this unit as a revision for yourself and scan through the text. If you score less than 80% then work through the text and re-test yourself at the end by using this same test.

The answers are at the end of the unit.

1 What are the names of the three main types of radiation emitted by radioactive materials? **(3)**
2 What does each type of radiation consist of? **(6)**
3 What happens to the atomic number and mass number of an atom when each type of radiation is emitted? **(8)**
4 Write equations for the emission of (a) an α-particle by fermium-253, atomic number 100 and (b) a β-particle by sodium-24, atomic number 11. **(6)**
5 The half-life of a newly discovered isotope is five years. If you weigh out 8 mg of the pure element and leave it for twenty years how much of the pure element will be left? **(3)**

26 Marks (80% = 21)

1.4.1 The Discovery of Radioactivity

There are often ultraviolet lights used in disco clubs. When they are turned on some clothing glows really brightly. That's because the clothes have been washed in detergents that contain chemicals which **fluoresce**; they absorb ultraviolet radiation and re-emit it as visible light and this makes it look brighter. That's why your washing looks 'whiter than white'. If you happen to be drinking gin and tonic in the club when the UV is switched on, you may notice your drink glowing; that's because tonic water contains quinine and quinine fluoresces. If you've got false teeth, though, don't open your mouth at the club; the plastic that some false teeth are made from fluoresces green in UV.

A French chemist called Henri Becquerel was interested in fluorescence. X-Rays had just been discovered in the late nineteenth century and Becquerel wondered whether any fluorescent substances gave off X-rays as well as light. To find out, he used a fluorescent compound of uranium. He put a photographic film in a light-proof envelope and put the uranium compound on top of that in the sunlight. After a while he developed the film; he found that it was fogged. Obviously he thought X-rays were being produced by fluorescence of the compound, passing through the light-proof wrapping and affecting the film.

During one sunless period, Becquerel put his envelopes, still with the uranium compound on them, away in a drawer to wait for good weather. After a couple of days with no break in the clouds he decided to develop the film anyway. When he did so, he was astonished to find that they were just as much affected as if there had been strong fluorescence taking place. Obviously there must be some **other** kind of radiation being produced by the compound which was independent of fluorescence or X-rays. Experiments soon showed that it was the uranium that was producing it.

One of the scientists who set to work to investigate this new kind of radiation, about a hundred years ago, was Marie Curie. She coined the word 'radioactivity' to describe the ability of uranium and other elements to give out radiation which could affect photographic plates. We can see that it is a confusing title, as it has nothing to do with 'radios' but a lot to do with radiation; it would have been better called 'radiation activity'. But what could this 'radiation' be? Was it a new type of radiation?

Marie Curie won two Nobel Prizes (only two other people have ever done this). The first was in 1903, with her husband Pierre and Henri Becquerel, for Physics; the second in 1911, for Chemistry.

1.4.2 Types of Radiation

It was soon found that this new type of radiation was even more penetrating than X-rays. It was given the name gamma (γ) radiation. It was also found that elements can emit other kinds of radiation as well. The other two most important are alpha (α) and beta (β) radiation.

What is the difference between these types of radiation?

Gamma (γ) radiation is electromagnetic radiation like light, radio waves and X-rays, but with much higher energy; to put it another way, it has higher frequency or shorter wavelength than light, radio or X-rays. Although Marie Curie didn't know it, it is also much more damaging than the other types of electromagnetic radiation; it penetrates further into living matter and damages cell material. Marie Curie died of leukaemia – almost certainly a result of long exposure to radiation.

The other two main types of radiation are much less penetrating than gamma rays. They are different from gamma in that they are not electro-magnetic radiation at all, but consist of sub-atomic particles. The more penetrating of the two is beta radiation, because it is made up of smaller particles than alpha radiation.

Another difference is that the path of these two types of radiation is affected differently by electrical and magnetic fields (see Figure 1.4.1).

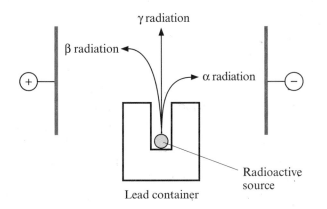

Figure 1.4.1
Effect of electrical and magnetic fields on radiation

Alpha (α) radiation is attracted towards the cathode or negative pole. This is because it is made up of positive particles. Beta (β) radiation, on the other hand, is attracted towards the positive pole, or anode, because it is made up of negative particles.

Beta radiation is more penetrating than alpha because it is made up of much smaller particles; beta particles stand a much smaller chance of interacting with the atoms of a material as they pass through it, and therefore stand a much smaller chance of being stopped. Beta particles are roughly one eight-thousandth the mass of alpha particles; in fact, beta particles are nothing more or less than electrons. Alpha particles, on the other hand, are made up of two protons and two neutrons; they are the same as helium nuclei (Figure 1.4.2).

Figure 1.4.2
α-Particle $^4_2$He, helium nucleus

1.4.3 Transmutation of Elements

The radiation given off by atoms must be coming from somewhere. It was later realised that it was coming from the nucleus. Now, if the nucleus of an atom gives out an alpha particle, it loses two protons; if it loses two protons its atomic number goes down by 2; it also loses two neutrons, so its mass number goes down by 4 (2 protons *and* 2 neutrons). The most important thing is the change in atomic number, because the number of protons is what determines what element the atom is. In other words, when an element emits an alpha particle, it changes to a different element. It is **transmuted**. This can be shown in equations of the type:

$$^{238}_{92}\text{U} \xrightarrow{\;\alpha\text{-emission}\;} {}^{234}_{90}\text{Th} + {}^4_2\text{He} \text{ (the } \alpha\text{-particle)}$$

The uranium atom's nucleus loses an alpha particle; its atomic number goes down from 92 to 90, so it becomes an atom of thorium; the mass number goes down from 238 to 234 because it has lost 2 neutrons and 2 protons.

What about emission of beta particles? Beta particles are electrons and electrons are in orbits round the nucleus, so surely the loss of an electron cannot affect the nucleus? It does, though. For example, the thorium atom produced in the reaction above is radioactive. It gives out beta radiation and when it does, the result is:

$$^{234}_{90}\text{Th} \xrightarrow{\ \beta\text{-emission}\ } {}^{234}_{91}\text{Pa} + {}^{\ 0}_{-1}\text{e} \ (\text{the } \beta\text{-particle})$$

The atomic number goes **up** by 1 to 91; the thorium atom changes to an atom of protactinium. The mass number doesn't change at all, because electrons have virtually zero mass.

How does the atomic number go up? The nucleus contains only protons and neutrons, or at least that's what we've said so far. Suppose that neutrons aren't simple particles. Suppose that they are actually made up of one electron and one proton; their mass will be virtually the same as the mass of a proton, because electrons have hardly any mass, and their charge will be zero because the negative of the electron will cancel out the positive of the proton.

If we assume that this picture is right, we can see how the loss of an electron **from the nucleus** will affect the atomic number: a neutron loses an electron and changes into a proton; the number of protons in the nucleus (*i.e.* the atomic number) goes up by 1:

$$n = (p^+ + e^-) \rightarrow \text{proton } p^+ \ ({}^{1}_{1}\text{H}) + {}^{\ 0}_{-1}\text{e} \ (\text{electron})$$

Notice that in this sort of equation, just as in maths, each side of the equation must balance. In this one, the sum of the mass numbers on the left is 1 and the sum of the mass numbers on the right is 1 (1 + 0). Similarly, the sum of the atomic numbers on the left is 0 and the sum of the atomic numbers on the right is 0 [1 + (−1)].

If you look back at the equation at the top of the page, you'll see that the same is true. The sum of the mass numbers on the left is 234 and the sum of the mass numbers on the right is 234 (234 + 0). Similarly, the sum of the atomic numbers on the left is 90 and the sum of the atomic numbers on the right is 90 [91 + (−1)].

If this idea of nuclear electrons is right, there is no such thing as a neutron. The electrons in the nucleus switch from proton to proton, acting like a kind of 'negative cement' holding the particles in the nucleus together and converting protons into neutrons and *vice versa*. The ratio between neutrons and protons in the nucleus would be important in determining the stability of the nucleus. This does seem to be the case as far as we can deduce from the evidence, but we will not really know until we are able to 'see' the nucleus using some yet undiscovered technique. That nucleus seems a mysterious place, and we will hear a lot more about it as researchers are able to unravel exactly what is going on in there. Of course, we are all made up of atoms, billions of them.

One thing you might have noticed was the use of the word 'transmuted' meaning the inter-conversion of one element into another. The ancient 'alchemists' were always trying to do this, their main object was to transmute cheap lead into gold. But they never achieved it. Transmutation of elements was going on all around them if they only knew it, but not lead into gold!!!. Well, not yet.

1.4.4 Half-life

Most carbon atoms have six protons and six neutrons in their nuclei. They are perfectly stable and don't decay, or split up, or emit radiation of any kind. This isotope of carbon is called 'carbon-12', because its mass number is 12. Another isotope of carbon is carbon-14, which has six protons and eight neutrons. This isotope is radioactive; it does undergo radioactive decay giving off beta particles and forming nitrogen-14. Remember that **isotopes** are atoms having the same atomic number (*i.e.* the same number of protons) but different mass numbers (*i.e.* different numbers of neutrons) – see Section 1.3.4.

If you were able to observe a single atom of carbon-14, you would know that it would eventually decay, but there would be no way of knowing exactly when it would happen. If you take billions of carbon-14 atoms, you can say nothing about when any individual nucleus will emit radiation, but you can say that after 5770 years half of them will have broken down. This is what is called the **half-life** of carbon-14. Different radioactive isotopes have different half-lives. The half-life of uranium-238, for example, is 4.5 billion years, and that of uranium-235 is 710 million years.

The present best guess at the age of the Earth is about 4.6 billion years. That means that, since the half-life of uranium 238 is 4.5 billion years, there must be about half as much uranium 238 left as there was when the Earth was first formed. The half-life of carbon-14, on the other hand, is only 5770 years, which means there ought to be hardly any left at all. The reason there is still some carbon-14 is because of cosmic rays. Cosmic rays contain high energy neutron particles that hit the Earth from outer space; as you read this, some of these particles will be passing through your body. Some of these neutrons collide with nitrogen-14 atoms in the air and convert them to carbon-14 with the emission of a proton. So carbon-14 is being produced all the time, and some of it is decaying all the time by beta decay back to nitrogen, which means that the level of carbon-14 in the atmosphere stays more or less constant:

$$^{14}_{7}\text{N} + ^{1}_{0}\text{n} \rightarrow ^{14}_{6}\text{C} + ^{1}_{1}\text{H}$$

$$^{14}_{6}\text{C} \rightarrow ^{14}_{7}\text{N} + ^{0}_{-1}\text{e}$$

Carbon dioxide containing this tiny proportion of carbon-14 is taken up by plants during photosynthesis and is incorporated into their cells. Once the plant dies, carbon-14 will stop being taken in and so the amount of carbon in the plant will start to decay. Suppose you dig up a piece of wood, extract the carbon and measure the amount of radioactivity. If you find that the level of carbon-14 activity is only about half of what it is today, then the tree from which the wood comes must have died about one half-life ago, *i.e.* about 5770 years. In fact, the way the activity will decrease is as follows:

in 5770 years	half as active
in 11 540 years	a quarter as active
in 17 310 years	an eighth as active
in 23 080 years	a sixteenth as active

and so on. Of course, it isn't quite as simple as this, but the general idea is true. For every half-life that goes past, the amount of the isotope, and hence its activity, is halved. This means that you can get an idea of the age of artifacts made of organic material, and of human remains, since they all contain carbon. This method is called 'carbon dating'.

For dating rocks, geologists have used the uranium-238 isotope, which decays to lead-206 *via* a number of intermediate products. The ratio of these two isotopes in a particular rock can be used to estimate the age of the rock.

Medical science also uses small traces of safe radioactive materials to track the passage of materials around a body or plant. Radio-thallium-201 is used to assess the damage to heart muscles after a person has had a heart attack. Radioactive iodine-131 is used in the treatment of thyroid disorders.

Some brands of smoke detectors contain a very small amount of the radioactive element americium.

Elements with long half-lives are produced as waste by nuclear power stations. These have to be stored until the activity has fallen to a safe level. Some of the material can be used in 'reprocessing plants' but no one likes these on their doorstep. All energy problems in life have their 'fors' and 'againsts'. Radioactive materials are all around us all the time, as they occur naturally in rocks. The problem comes when we concentrate them for use for our benefit and then cannot easily get rid of the waste.

Answers to diagnostic test

If you score less than 80%, then work through the text and re-test yourself at the end using this same test. If you still get a low score then re-work the unit at a later date.

1 Alpha, beta and gamma   (1 mark for each)   **(3)**

2 Alpha = helium nuclei (2 protons and 2 neutrons)   (3 marks)
   Beta = electrons from the nucleus   (2 marks)
   Gamma = electromagnetic radiation   (1 mark)   **(6)**

3 Alpha   Atomic number goes <u>down</u> by <u>2</u>   Mass number goes <u>down</u> by <u>4</u>
   Beta   Atomic number goes <u>up</u> by <u>1</u>   Mass number <u>does not change</u>
   Gamma   <u>Neither change</u>   (1 mark for each underlined)
                                                                                              (8)

4 a) $^{253}_{100}\text{Fm} \rightarrow \ ^{249}_{98}\text{Cf} + \ ^{4}_{2}\text{He}$ (1 mark for each species correct)   **(3)**
   b) $^{24}_{11}\text{Na} \rightarrow \ ^{24}_{12}\text{Mg} + \ ^{0}_{-1}\text{e}$ (1 mark for each species correct)   **(3)**

5 20 years = 4 half-lives   (1 mark)
   Amount left = $8 \times \frac{1}{2} \times \frac{1}{2} \times \frac{1}{2} \times \frac{1}{2}$ mg   (1 mark)
                       = 0.5 mg   (1 mark)   **(3)**

26 Marks   (80% = 21)

FURTHER QUESTIONS (NO ANSWERS GIVEN)

1 State what you think will be formed in each of the following radioactive decay reactions:

   a) Beta decay of hydrogen-3 (tritium).
   b) Beta decay followed by alpha decay of lithium-8.
   c) Beta decay of phosphorus-32.

2 A smoke detector contains a very small trace of americium-241 which decays by alpha release. What will be formed?

   a) The complete decay sequence of the americium-241 goes *via* the following scheme. Write the chart of the sequence.
      Am-241 decays by α, α, β, α, α, β, α, α, α, β, α, β.
   b) Can you work out the intermediate elements by looking at the periodic table?

3 There are some fascinating areas of radioactivity that can be followed up if you have an interest. Such areas could include . . . what is meant by fission and fusion?

4 Chadwick bombarded beryllium-9 (atomic number 4) with alpha particles and produced neutral particles called neutrons. Write the nuclear equation for this. What else would be produced?

5 Complete the following table.

Element	At. no.	At. mass	No. of protons	No. of neutrons	Electronic structure	Metal/ non-metal	Mass number of other isotopes
Helium		4					
	6	12					14
Sodium		23					
			12	12			
		31			2:8:5	Non-metal	
Krypton		84					
Vanadium	23			28			
			30	35	2:8:8:10:2	Metal	
Strontium		88					
	47	108				Metal	
Platinum		195					

6 What is X in the following sequence?

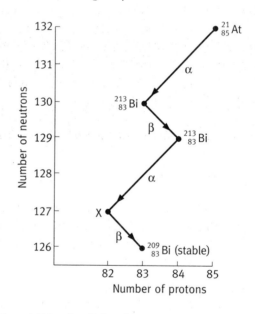

7 What are Y and Z in the following two-step decay sequence:

$$^{14}_{7}\text{N} + ^{1}_{0}\text{n} \rightarrow ^{14}_{6}\text{C} + \text{Y}$$

$$^{14}_{6}\text{C} \rightarrow \text{Z} + ^{0}_{-1}\text{e}$$

Unit 1.5
Electrons

Starter check

What you need to know before you start the contents of this unit.

You need to understand about the electronic structure of the atom.

Aims **By the end of this unit you should understand that:**

- Orbits are energy levels.
- Electrons can move between energy levels.
- Movements between energy levels involve absorption or emission of energy.
- Energy is emitted or absorbed in the form of electromagnetic radiation.
- Each electron move absorbs or emits energy of a specific wavelength.
- Orbits are split into orbitals.

Diagnostic test

Try this test at the start of the unit. If you score more than 80%, then use this unit as a revision for yourself and scan through the text. If you score less than 80% then work through the text and re-test yourself at the end by using this same test.

The answers are at the end of the unit.

1 What is an absorption spectrum? **(3)**
2 Which have the greater energy, electrons close to or further from the nucleus? **(1)**
3 Why can atoms only absorb light of characteristic wavelength? **(4)**
4 Which energy level is furthest from the nucleus, $n = 1$ or $n = 4$? **(1)**
5 What name is given to orbitals for which (a) $l = 0$ (b) $l = 1$ (c) $l = 2$ (d) $l = 3$? **(4)**
6 How many p-orbitals are there in each energy level? **(2)**
7 How many d-orbitals are there in each energy level? **(3)**
8 How many f-orbitals are there in each energy level? **(4)**
9 What is meant by 'ground state configuration'? **(1)**
10 What is the ground state configuration of an element with (a) 10 and (b) 15 electrons? **(8)**

31 Marks (80% = 25)

1.5.1 Electrons in Atoms

Figure 1.5.1
A hydrogen atom

When the idea of atoms being made up of a positive nucleus surrounded by negative electrons was first put forward, the idea was that the electrons orbited the nucleus just as planets orbit the Sun. Just as the force of gravity keeps the planets in their orbits and keeps them from flying off into space, so the force of attraction between the positive nucleus and the negative electrons keeps them in their orbits.

The simplest atoms are hydrogen atoms (Figure 1.5.1); they have just one electron.

Imagine twirling a weight round on the end of a piece of elastic; if you twirl more quickly, the elastic will stretch and the weight will get further away from your hand. Suppose that the electron takes in some energy from somewhere. It will start to move more quickly. As it starts to move more quickly, it will move further away from the nucleus. The further the orbit is from the nucleus, the more the energy associated with the electrons in it. Now imagine that you shine some white light through a sample of hydrogen. White light is made up of light of lots of different wavelengths. Light is a form of energy – the shorter the wavelength, the greater the energy of the light. Red light has the longest wavelength (least energy) of visible light and violet has the shortest wavelength. If the electron absorbs red light, it will move a little away from the nucleus; if it absorbs violet light, it will move further away. There seems no obvious reason why an electron should absorb light of one wavelength rather than another.

As white light passes through the hydrogen gas, the electrons in some atoms would absorb red, those in others orange or yellow or green or blue or violet. When the white light reached the other side of the sample of hydrogen, some of each colour would have been absorbed; the intensity of each colour would have been reduced, but they will all still be there (unless there was so much hydrogen that all the light has been absorbed). So if you were to pass the light through a prism and shine it on to a screen, you would expect to see a full spectrum from red to violet as in Figure 1.5.2.

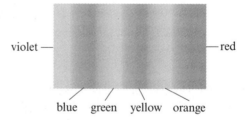

violet ——   —— red

blue   green   yellow   orange

Figure 1.5.2
The visible spectrum

In fact, that isn't what you see. Most of the colours come through the sample totally unchanged. Some of the wavelengths, on the other hand, have totally disappeared. There are gaps in the spectrum where these wavelengths have been taken out.

1.5.2 Absorption Spectra

At noted above, if you pass white light through hydrogen and then view the spectrum, you will find that most of the different colours are still there; they haven't been absorbed at all. But some very specific wavelengths have been totally absorbed; they don't come through, and are represented by black lines in the spectrum shown in Figure 1.5.3.

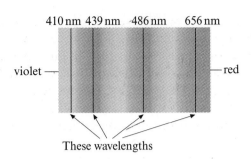

violet —

— red

Figure 1.5.3
The hydrogen absorption spectrum

The same sort of thing happens with all other elements when made into vapours. They absorb a few specific wavelengths and allow the rest to pass through. This sort of spectrum is called an **absorption spectrum**. Each element absorbs a particular set of wavelengths. You can identify the different elements by seeing which wavelengths are absorbed, *i.e.* by seeing what the absorption spectrum looks like. This sort of analysis is called **spectroscopy**. The instruments used to carry it out are called **spectroscopes**.

1.5.3 Why Do Elements Only Absorb Certain Wavelengths?

Because only certain orbits exist within an atom, only specific wavelengths of light are absorbed. Electrons can only absorb light with a wavelength that exactly corresponds to the energy change involved in each movement of the electrons between these orbits (Figure 1.5.4).

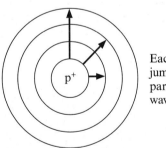

Each of these jumps absorbs a particular wavelength

Figure 1.5.4
Specific wavelengths of light are absorbed

The electrons in each of these orbits possess a particular amount of energy, so the orbits are called **energy levels**. The word 'quantum' means 'a particular amount', so movement to each of the energy levels involves a specific quantum of energy; each energy is thus given a **principal quantum number**, represented by the letter 'n' (Figure 1.5.5).

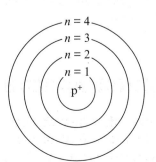

Figure 1.5.5
Energy levels for the principal quantum number, n

1.5.4 More Accurate Spectroscopes, and Quantum Numbers

Figure 1.5.6
Sub-levels

The earliest spectroscopes were not as good as they might have been – the spectra they showed were a little blurred; we say that their resolution wasn't too good. As time went on, spectroscopes improved, their resolution got better, and it was realised that some of the lines in the spectra of the elements were not single lines at all; they were actually what are called doublets – two lines placed very close together. Some of them were triplets. How can this be explained?

Assume that in the second energy level ($n = 2$), there are two types of sub-level possible, one of them with a circular orbit and the other one elliptical, as shown in Figure 1.5.6.

The $n = 2$ energy level contains two different types of sub-level; the subdivisions are called 'orbitals'. The electrons in each orbital will have **slightly** different energy. A jump from $n = 1$ to $n = 2$ (circular) will absorb a slightly different wavelength to a jump from $n = 1$ to $n = 2$ (elliptical). There will be two absorption lines very close together in the spectrum.

The number of doublets, triplets, *etc.* in the spectra of various elements can be explained nicely if you assume that there is only one type of sub-level in the $n = 1$ energy level, two in the $n = 2$, three in the $n = 3$, and so on. A second quantum number, called the **azimuthal quantum number**, and given the symbol 'l', is assigned to each sub-level within an energy level. If 'l' is allowed to have the values 0 to $(n - 1)$ for each energy level, you get the right number of sub-levels (Figure 1.5.7),

	Values of n	*Values of* l [*or* (n−1)]
1st E level	1	0
2nd E level	2	0
	2	1
3rd E level	3	0
	3	1
	3	2

Figure 1.5.7
Values of *n* and *l* for each energy level

and so on.

Orbitals for which 'l' = 0 are deemed to be circular and are called 's'-orbitals. From Figure 1.5.7 you can see that the first energy level only contains an s-orbital. It is called the 1s orbital because 'l' = 0 and it is in energy level 1. Similarly there is a 2s-, a 3s- and a 4s-orbital.

Orbitals for which 'l' = 1 are elliptical and are called 'p'-orbitals. There are no p-orbitals in the first energy level, because when n is 1, 'l' can only be 0. There is a 2p-, a 3p- and a 4p-orbital, and so on.

Orbitals for which 'l' = 2 are called 'd'-orbitals; orbitals for which 'l' = 3 are called 'f'-orbitals. From this the information in Figure 1.5.7 can be expanded, to give Figure 1.5.8.

This was all fine, until it was realised that some of the single sharp lines in the spectra split when the sample was exposed to a strong magnetic field. This can be explained by assuming that the elliptical orbitals are actually groups of orbitals which align themselves differently in a magnetic field. Thus there are **three** p-orbitals, **five** d-orbitals. This is best explained by assigning a third, **magnetic quantum number**, m, and allowing it to have values $+l$ to $-l$: attracted or repelled by the magnetic field. The shapes of p- and d-orbitals are shown in Figure 1.5.9.

	Values of n	Values of l	Sub-level
1st E level	1	0	1s
2nd E level	2	0	2s
	2	1	2p
3rd E level	3	0	3s
	3	1	3p
	3	2	3d
4th E level	4	0	4s
	4	1	4p
	4	2	4d
	4	3	4f

Figure 1.5.8
Names of sub-levels for different n and l

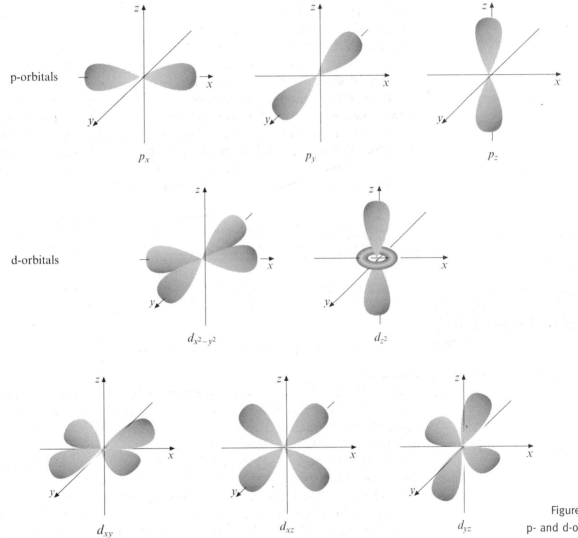

Figure 1.5.9
p- and d-orbitals

Just two more things to add now. The first is that electrons spin in their orbitals; some spin clockwise, and some spin anti-clockwise. The second is that each orbital can hold just two electrons – one electron spinning clockwise and the other anti-clockwise.

The way the electrons are arranged in particular elements is called the **electronic configuration** of the element. It is common to give the electronic configuration when the electrons are in the lowest energy orbitals available to them; this is called the **ground state configuration** for the element.

Hydrogen has just one electron. It goes into the first energy level. There is only a 1s-orbital in the first energy level, so hydrogen's ground state configuration is $1s^1$, *i.e.* one electron in the 1s orbital. Helium has 2 electrons; the 1s-orbital can hold 2 electrons, each spinning in the opposite direction, so helium's ground state configuration is $1s^2$.

Lithium has 3 electrons. The 1s-orbital is full, so the next electron goes into the 2s-orbital.

Beryllium has 4 electrons. The extra electron fills the 2s, so we have:

Li $1s^2 2s^1$ Be $1s^2 2s^2$

The 1s- and 2s-orbitals are now full, so the next electron goes into the 2p-orbitals. There are three 2p-orbitals, so the next six elements have the following ground state configurations:

B $1s^2 2s^2 2p^1$ O $1s^2 2s^2 2p^4$
C $1s^2 2s^2 2p^2$ F $1s^2 2s^2 2p^5$
N $1s^2 2s^2 2p^3$ Ne $1s^2 2s^2 2p^6$

Similarly, the ground state configurations of the next eight elements are as follows:

Na $1s^2 2s^2 2p^6 3s^1$ P $1s^2 2s^2 2p^6 3s^2 3p^3$
Mg $1s^2 2s^2 2p^6 3s^2$ S $1s^2 2s^2 2p^6 3s^2 3p^4$
Al $1s^2 2s^2 2p^6 3s^2 3p^1$ Cl $1s^2 2s^2 2p^6 3s^2 3p^5$
Si $1s^2 2s^2 2p^6 3s^2 3p^2$ Ar $1s^2 2s^2 2p^6 3s^2 3p^6$

The next electron should go into the 3d-orbitals, but it doesn't. Instead it goes into the 4s. This is because the energy level of the 4s orbital is less than the energy level of the 3d-orbital once there are enough protons in the nucleus, and in general electrons will go into the lowest energy level available. There is still a pattern for the filling of the orbitals, however, which is best shown using the diagram in Figure 1.5.10.

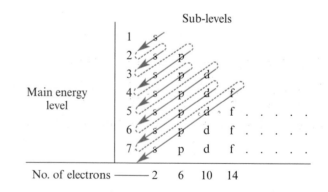

Figure 1.5.10
Filling pattern for orbitals

Thus, the ground state configurations of the next two elements are:

K $1s^2 2s^2 2p^6 3s^2 3p^6 4s^1$ Ca $1s^2 2s^2 2p^6 3s^2 3p^6 4s^2$

Sc $1s^2 2s^2 2p^6 3s^2 3p^6 3d^1 4s^2$

From scandium to zinc the 3d-shell begins to fill up until it is full with 10 electrons, then the next most favourable energy level starts to fill up, *i.e.* the 4p-orbital.

Answers to diagnostic test

If you score less than 80%, then work through the text and re-test yourself at the end using this same test. If you still get a low score then re-work the unit at a later date.

1 Spectrum produced by <u>passing light through</u> a substance and noting which <u>wavelengths</u> are <u>absorbed</u>.
   (1 mark for each term underlined)   **(3)**

2 Further from.   **(1)**

3 Atoms contain electron energy levels.
   Absorption of light involves electrons moving to higher energy levels.
   Such electron jumps involve specific energy changes.
   Therefore the absorbed light is of specific wavelengths.
   (1 mark for each correct sentence)   **(4)**

4 $n = 4$ **(1)**

5 a) s-orbital   (1 mark)
   b) p-orbital   (1 mark)
   c) d-orbital   (1 mark)
   d) f-orbital   (1 mark)   **(4)**

6 None in $n = 1$ (1 mark); 3 in the rest (1 mark)   **(2)**

7 None in $n = 1$ or $n = 2$ (2 marks); 5 in the rest (1 mark)   **(3)**

8 None in $n = 1$, $n = 2$ or $n = 3$ (3 marks); 7 in the rest
   (1 mark)   **(4)**

9 The configuration in which each electron occupies the lowest energy level available.   **(1)**

10 a) $1s^2 2s^2 2p^6$ **(3)**
   b) $1s^2 2s^2 2p^6 3s^2 3p^3$ **(5)**

31 Marks   (80% = 25)

FURTHER QUESTIONS (NO ANSWERS GIVEN)

1 What is the sequence for filling up the electron shells using the s-, p-, d-orbitals up to element number 38?

2 When do the 4f-orbitals begin to fill up? Why are there no 2f- or 3f-shells?

Module 2

Chemical Bonding in Materials

Unit 2.1
Atoms and Ions

Starter check

What you need to know before you start the contents of this unit.

You should have covered the material in Module 1.

Aims **By the end of this unit you should understand that:**

- When atoms react they usually form a noble gas structure.
- Electrons can move between atoms
- When atoms lose or gain electrons, they are no longer neutral.
- When atoms lose electrons they become positive ions.
- When atoms gain electrons they become negative ions.
- Positive and negative ions attract each other to form compounds.

Diagnostic test

Try this test at the start of the unit. If you score more than 80%, then use this unit as a revision for yourself and scan through the text. If you score less than 80% then work through the text and re-test yourself at the end by using the same test.

The answers are at the end of the unit.

1 What are the electrons in atoms trying to achieve when they combine with each other? **(1)**

2 What sort of ions would you expect an element in Group 2 to produce? **(2)**

3 What sort of ions would you expect an element in Group 6 to produce? **(2)**

4 Draw dot and cross diagrams to represent electrons to show how you would expect atoms for the following elements to combine, what ions would be produced, and what the formulae of the resulting compounds would be.

a) Lithium and fluorine; **(5)**
b) magnesium and fluorine; **(5)**
c) magnesium and oxygen; **(5)**
d) lithium and nitrogen; **(5)**
e) magnesium and nitrogen. **(5)**

30 Marks   (80% = 24)

2.1.1 Why Atoms Combine

Each element is made up of atoms all of the same type. Atoms are made up of the much smaller particles, of protons, neutrons and electrons. Atoms are neutral because they have the same number of negative electrons as they have positive protons.

The big questions that arise are . . . 'Why do atoms react with each other? Why don't they just stay as they are – as separate atoms?'

Some atoms do just that, stay as they are. The Noble Gases in Group 0 of the periodic table, don't react with much at all. They exist as single atoms and they stay that way. They are, so to speak, 'happy with their lot'. They don't need to join up with anything else to gain the stability of a complete outer electron shell. They already have their outer electron shell full and complete.

What happens in a chemical reaction?

Before any atom can react with another one, they have got to meet. An atom in one corner of the room cannot react with one in the other corner. They have to collide with each other. When they collide, it's obviously the outside of the atoms that come into contact with each other. So maybe the place to look to explain why atoms do or don't react, and why they react the way they do, is the outside of the atom.

What is on the outside of the atom? The electrons. In fact we don't need (mostly) to look any further than just the electrons on the very outside of the atom – the electrons in the outermost energy level – when considering the chemical reactions of elements.

What is there in common about the outermost energy levels of atoms of the noble gases? Apart from helium, they all have eight electrons in the outermost energy level. Helium's outermost energy level cannot hold eight electrons (eight electrons are called an **octet**), because it is the first energy level and much smaller and so can only hold two. The configuration, an octet in the outermost energy level, must be a particularly stable arrangement because the elements that have it don't try to change it, they are generally unreactive or inert. Atoms that don't have the noble gas configuration try to get it. How can they do that?

2.1.2 How Atoms Combine

Let us start by looking at the periodic table and elements in Group 1 – the Alkali Metals. All of these elements have got one electron in their outermost energy level. (You can see how useful the periodic classification is – we can talk about all the elements in Group 1 at the same time, because their electronic configurations are so similar, and particularly because their outside shells contain the same number of electrons, namely 1.) Let us take sodium as our example. Its ground state configuration is $1s^2 2s^2 2p^6 3s^1$. We can show this using diagrams as in Figure 2.1.1.

If a sodium atom can manage to get rid of an electron, its configuration will be $1s^2 2s^2 2p^6$, which is the same as the configuration of neon, a noble gas. So what's the difference?

A neon atom has 10 electrons and 10 protons, so it is neutral; the sodium atom did have 11 electrons, but it has lost one so that it also has 10 electrons – but it still has 11 protons. It has one more positive charge

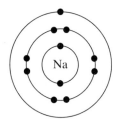

Figure 2.1.1
The sodium atom

than negative charges, so it is no longer neutral; it has an overall positive charge of +1. Atoms are neutral; this particle is positively charged, so instead of calling it a sodium atom, we call it a sodium **ion** (Figure 2.1.2).

Now let us look at the elements in Group 7 – the Halogens. All these elements have got seven electrons in their outermost energy level. Let us take chlorine as our example. Its ground state configuration is $1s^2 2s^2 2p^6 3s^2 3p^5$. We can show this in Figure 2.1.3.

If a chlorine atom can manage to get hold of an extra electron from somewhere, its configuration will be $1s^2 2s^2 2p^6 3s^2 3p^6$, which is the same as the configuration of argon, a noble gas. So what's the difference?

An argon atom has 18 electrons and 18 protons, so it is neutral; the chlorine atom did have 17 electrons, but it has gained one so it also has 18 electrons, but it still has 17 protons. The chlorine now has one more negative charge than positive charges, so it is no longer neutral; it has an overall negative charge of −1. Atoms are neutral; this new particle is negatively charged, so instead of calling it a chlorine atom, we call it a chloride ion (Figure 2.1.4).

Now suppose that a sodium and a chlorine atom collide with each other. The sodium can get rid of one electron by giving it to the chlorine atom: both now have got the desired 'noble gas' configuration (Figure 2.1.5).

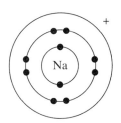

Figure 2.1.2
The sodium ion

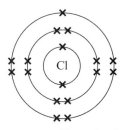

Figure 2.1.3
The chlorine atom

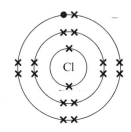

Figure 2.1.4
The chloride ion

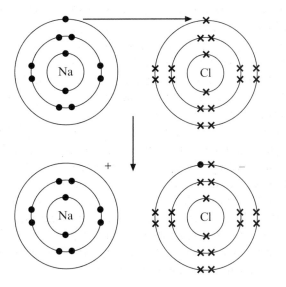

Figure 2.1.5
Sodium chloride

This type of diagram is usually called a 'dot and cross' diagram because the electrons from the different atoms are shown as dots and crosses (although, of course, there is no real difference between the electrons of different atoms).

Just look at what is produced: one positive sodium ion and one negative chloride ion. In any real sample of sodium and chlorine, of course, there would be billions of ions, but for every one sodium ion there would be one chloride ion. They would attract each other because they have opposite charges to form a compound, sodium chloride. Its formula is NaCl because the ratio of sodium ions to chloride ions is 1:1. See also Section 2.3.1 for further details.

Elements in Group 2 have two electrons in their outermost energy level, so they tend to lose two electrons and form ions with a double positive charge. Similarly, elements in Group 3 have three electrons in their outermost energy level, so they tend to lose three electrons and form ions with a triple positive charge, although it's much harder to lose three electrons than it is to lose one or two.

In the same way, elements in Group 6 tend to take in two electrons to get a stable octet, forming ions with a double negative charge, and elements in Group 5 tend to take in three electrons to get a stable octet, forming ions with a triple negative charge.

Let us have a look at what happens when an element from Group 2 combines with an element from Group 6:

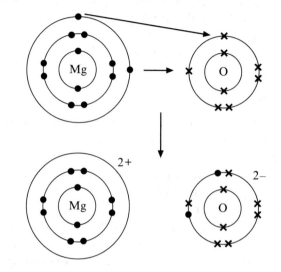

Figure 2.1.6
Reaction between magnesium
and oxygen

As you can see from Figure 2.1.6, one atom of magnesium gives two electrons to one atom of oxygen, so the formula of the compound is MgO.

Now let us have a look at what happens when an element from Group 3 combines with an element from Group 6:

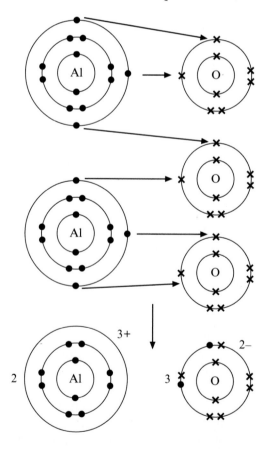

Figure 2.1.7
Reaction between aluminium
and oxygen

As you can see from Figure 2.1.7, two atoms of aluminium give six electrons to three atoms of oxygen, so the formula of the compound is Al_2O_3.

In general, you can assign numbers to the elements in Groups 1, 2, 3,

5, 6 and 7 that give the number of charges on the ions they form like this:

Group   1   2   3   4   5   6   7
          +1   +2   +3   0   −3   −2   −1

Once you know this, it's quite easy to work out the formulae of compounds formed by **ionic bonding** like this where electrons are **transferred** between atoms.

Notice that the elements on the left-hand side of the periodic table:

a) are metals; and
b) tend to react by forming positive ions.

The reverse is true for elements on the right-hand side of the table. The ones on the far right in Groups 6 and 7 are generally non-metals and tend to form negative ions.

Sometimes it is difficult for the elements in the centre of the table to know whether to gain or lose electrons, and they generally form a stable noble gas structure in their chemical reactions by using a mechanism other than ionic bonding. This is discussed in the next unit.

Note that there is no such thing as an individual ionic bond. Ionic compounds are made up of billions of ions held together by electrostatic attraction. See also Section 2.3.1.

Answers to diagnostic test

If you score less than 80%, then work through the text and re-test yourself at the end using this same test. If you still get a low score then re-work the unit at a later date.

1 Inert gas (or noble gas) structure.   **(1)**
2 Ions with <u>two</u> <u>positive</u> charges.   **(2)**
3 Ions with <u>two</u> <u>negative</u> charges.   **(2)**
4 Half a mark each for: correct structure of each atom and ion, correct charge on each ion, correct number of each atom, correct number of each ion.

a)

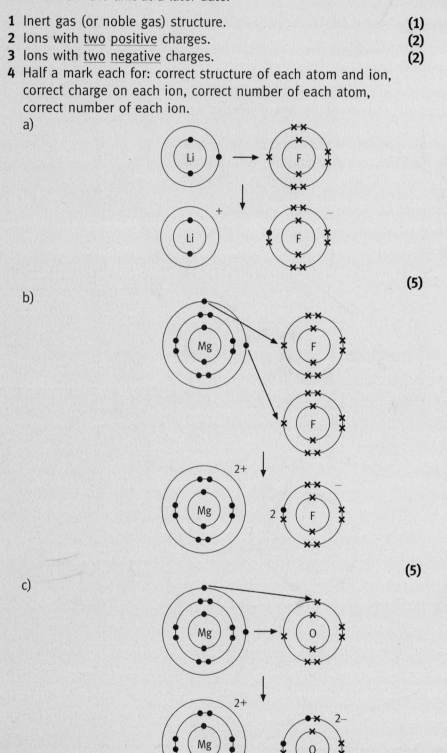

(5)

b)

(5)

c)

(5)

d)

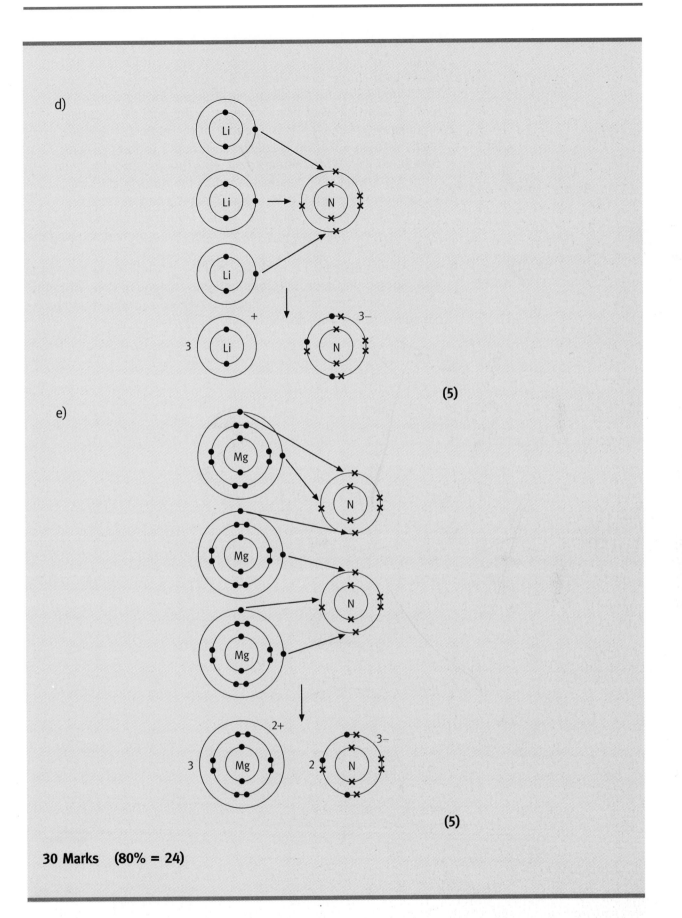

(5)

e)

(5)

30 Marks   (80% = 24)

FURTHER QUESTIONS (NO ANSWERS GIVEN)

1 An unknown element A is in Group 1 of the periodic table and reacts with an unknown element B in Group 6 to form an ionic compound AB. Write the electronic structure for the compound formed, showing what has happened to the outside electrons.

2 An unknown element X is in Group 2 of the periodic table and reacts with an unknown element Y in Group 7 to form an ionic compound XY_2. Write the electronic structure for the compound formed, showing what has happened to the outside electrons.

3 An unknown element E is in Group 3 of the periodic table and reacts with oxygen in Group 6 to form an ionic compound E_2O_3. Write the electronic structure for the compound formed showing the outside electrons.

4 An unknown element J is in Group 2 of the periodic table and reacts with nitrogen in Group 5 to form an ionic compound J_3N_2. Write the electronic structure for the compound formed, showing what has happened to the outside electrons.

Unit 2.2
Molecules and Sharing Electrons

Starter check

What you need to know before you start the contents of this unit.

The unit assumes that you are familiar with the terms and contents of Module 1.

Aims **By the end of this unit you should understand that:**

- When atoms react they usually form a noble gas structure.
- Atoms can share electrons.
- To share electrons, atoms have to stay together to form molecules.
- A shared pair of electrons constitutes a single covalent bond.
- Molecules are electrically neutral.

Diagnostic test

Try this test at the start of the unit. If you score more than 80%, then use this unit as a revision for yourself and scan through the text. If you score less than 80% then work through the text and re-test yourself at the end by using this same test.

The answers are at the end of the unit.

1 What is a triatomic molecule? (2)
2 Draw dot and cross diagrams to show the electron distribution in the following:

   a) a molecule of chlorine; (3)
   b) a molecule of hydrogen chloride; (4)
   c) a molecule of sulfur dichloride; (5)
   d) a molecule of carbon dioxide. (5)
3 Why are metals unlikely to form covalent bonds? (1)
4 Which of the molecules in Question 2 is likely to show no separation of charge and why? (2)

22 Marks (80% = 18)

2.2.1 Sharing Electrons

Imagine two hydrogen atoms colliding with each other. Each has one electron in its outermost energy level (configuration $1s^1$). Each would like to have a noble gas structure, in this case helium ($1s^2$), so each wants to gain an electron. The first hydrogen atom says to the second hydrogen atom, 'Give me an electron'. The second hydrogen atom, which is rather a vulgar atom, says to the first hydrogen atom, 'Nuts! You give me one!'

This high level conversation goes on for some time (several zillionths of a second). In human affairs these mutual demands would undoubtedly lead to fisticuffs, but in the world of atoms, possibly because they have no fists to cuff, it leads to co-operation. Eventually the atoms come up with the bright idea of sharing. Politicians take note!!! That way, they can at least each have a share in a noble gas configuration. This can be shown in the dot and cross electron diagram of Figure 2.2.1.

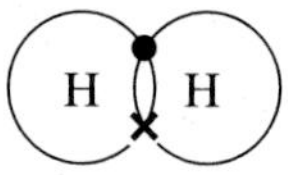

Figure 2.2.1
The hydrogen
molecule

2.2.2 Bonds

Obviously, once they've decided to share, they have to stick together; they are bonded. What is produced is a hydrogen molecule, which is made up of two hydrogen atoms. The two electrons they are sharing is what holds them together, or to put it another way, the electrons are the bond between the atoms. You can think of the two shared electrons as being between the two nuclei and attracting both. This type of bonding is called **covalent bonding** and the shared pair of electrons is called a covalent bond. In fact you very, very rarely find hydrogen atoms as single atoms; any normal sample of hydrogen is made up of hydrogen molecules, H_2. Hydrogen molecules are diatomic, that is they contain two atoms. That is why hydrogen is almost always written in chemical equations as H_2 rather than as just H.

Each hydrogen atom in a hydrogen molecule has one proton in its nucleus, and one electron in its half of the shared pair; that means that hydrogen molecules are electrically neutral. This is generally true – molecules are made up of atoms that have neither lost nor gained electrons, so molecules are neutral.

It would be rather inconvenient to have to draw a dot and cross diagram every time to describe the bonding in molecules, so a kind of shorthand is used. We simply use a line joining the two atoms to represent a shared pair of electrons or single covalent bond. The line represents one electron from each atom.

H–H

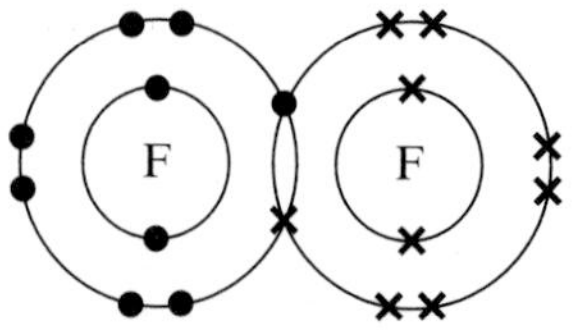

What about other elements? Do they share electrons as well? Think about fluorine: it is in Group 7, with a configuration of $1s^2 2s^2 2p^5$ – only one electron short of the neon configuration ($1s^2 2s^2 2p^6$). So each fluorine atom is looking for another electron, and when two of them meet they satisfy their requirements by sharing (Figure 2.2.2).

Similarly, oxygen is in Group 6, with a configuration of $1s^2 2s^2 2p^4$ – only two electrons short of the neon configuration ($1s^2 2s^2 2p^6$). So each oxygen atom is also looking for another two electrons, and when two of them meet they satisfy their requirements by sharing (Figure 2.2.3).

Notice that in this case each atom shares two electrons: there are two

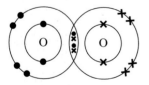

Figure 2.2.2
Fluorine molecule

Figure 2.2.3
Oxygen molecule

shared pairs, or a double covalent bond between the two atoms, O=O.

In Group 5, nitrogen has a configuration of $1s^2 2s^2 2p^3$ – three electrons short of the neon configuration ($1s^2 2s^2 2p^6$). So each nitrogen atom is looking for another three electrons, and when two of them meet they satisfy their requirements by sharing (Figure 2.2.4).

Notice that in this case each atom shares three electrons: there are three shared pairs, or a triple covalent bond between the two atoms, N≡N.

What we have looked at so far is the formation of covalent bonds between atoms of the same element, but they can also be formed between atoms of different elements. Take hydrogen and fluorine, for example. Both have atoms that are one electron short of a noble gas configuration, and they can share electrons in the same way as fluorine, oxygen and nitrogen do (Figure 2.2.5). Similarly, hydrogen and oxygen can bond to form water molecules, as shown in Figure 2.2.6, and hydrogen and nitrogen bond to form ammonia (Figure 2.2.7).

2.2.3 Equal and Unequal Shares

Notice that oxygen forms two covalent bonds whether it is bonding with another oxygen atom or with hydrogen. Similarly, nitrogen forms three covalent bonds whether it is bonding with another nitrogen atom or with hydrogen. There are differences, though. When an oxygen atom shares electrons with another oxygen atom, they each get a 'fair share' of the shared electrons. When hydrogen and oxygen share electrons, however, the oxygen atom takes more than its fair share of the shared electrons. We say that oxygen is more **electronegative** than hydrogen. This means that a water molecule isn't perfectly neutral (see Figure 2.2.8). The oxygen 'end' of the molecule has a slight negative charge, because it has more than its fair share of the shared pair, and the hydrogen 'ends' of the molecule have a slight positive charge for the same reason.

In fact, it is generally true that for any molecule that is made up of more than one type of atom from different elements there will be a slight separation of charge within that molecule. For the bond between two atoms of different elements A and B, if the two elements are similar and have similar electronegativities, then the compound AB would probably be covalent. If the values are greatly different (*e.g.* like elements at opposite ends of the periodic table), the bonding between A and B would be ionic. Atoms with slightly different values would tend to form covalent bonds between A and B, but the bond might be slightly distorted to cause one end of the bond (the more electronegative end) to draw electrons towards itself and so cause the bond to become **polar**, *i.e.* one end more negative than the other.

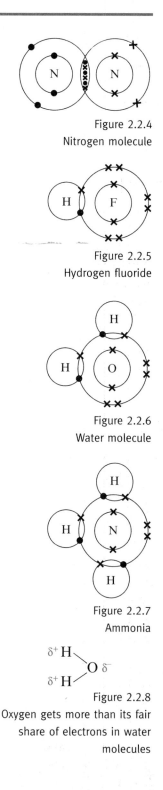

Figure 2.2.4
Nitrogen molecule

Figure 2.2.5
Hydrogen fluoride

Figure 2.2.6
Water molecule

Figure 2.2.7
Ammonia

Figure 2.2.8
Oxygen gets more than its fair share of electrons in water molecules

Which type of bond?

A $\overset{\bullet}{\underset{\times}{}}$ B   A and B are equally electronegative – **covalent** bond.

$^{\delta+}$A $\quad \overset{\bullet}{\underset{\times}{}}$ B$^{\delta-}$ B rather more electronegative than A – **polar covalent** bond.

$^{\oplus}$A $\quad \overset{\bullet}{\underset{\times}{}}B^{\ominus}$ B much more electronegative than A – **ionic** bond.

Further reading in this area can be done by looking up what is meant by 'electronegativities'.

One more thing. Metals tend to have only one, two or three electrons in their outermost energy levels. They are **not** likely, therefore, to be able to get a noble gas structure by sharing. If magnesium, for example, tries to share with chlorine, each chlorine atom can attain an inert gas structure, but the magnesium atom cannot (Figure 2.2.9).

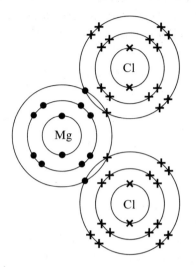

Figure 2.2.9
The impossible magnesium chloride structure

It can be seen that although the chlorine shells are full, the magnesium outer shell has not got a full octet of electrons, and so this structure is **not** the one that is formed in magnesium chloride.

As a general rule of thumb, if a compound contains a metal it will be ionic, and if it contains no metal it will be covalent.

Hydrogen can form covalent or ionic bonds . . . there is always one awkward customer! It is of course the smallest element in the periodic table, and so has some unusual properties.

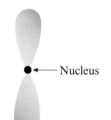

Figure 2.2.10
The p-orbital

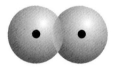

Figure 2.2.11
Overlap of s-orbitals in the hydrogen molecule

2.2.4 Shapes of Orbitals and Bonds

Orbitals are the regions in space that the electrons are said to frequent, and s-orbitals are spherical. p-Orbitals are shaped like dumb-bells, as shown in Figure 2.2.10.

You can think of covalent bonds as being the overlap between two orbitals, each of which has one electron in it. In hydrogen (configuration $1s^1$) molecules, for example, the bond can be pictured as in Figure 2.2.11. Similarly, hydrogen (configuration $1s^1$) bonds with oxygen (configuration $1s^2 2s^2 2p^4$) to form water (Figure 2.2.12).

The overlap represents a space where the shared electrons spend a lot of time, so it has increased negativity and holds the two positive nuclei together in their appropriate positions. When orbitals have a good overlap, the covalent bonding is called a **sigma (σ) bond**.

Sometimes each of two atoms has a partly filled p-orbital, and instead of overlapping end-to-end to form a sigma bond they overlap sideways in a less effective manner, to form what is called a **pi (π) bond** (Figure 2.2.13).

Figure 2.2.12
Orbital overlap in water

Figure 2.2.13
Sideways overlap of p-orbitals to form a π-bond

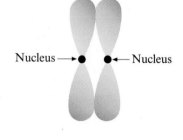

π-Bonding is particularly important when considering the properties, of organic compounds called alkenes, arenes and carbonyl compounds, to be discussed in Module 6. An example given here in Figure 2.2.14 is for ethene; it shows the sideways overlap of the p-orbitals, called the π-bond. The straight lines between the C and H are σ-bonds.

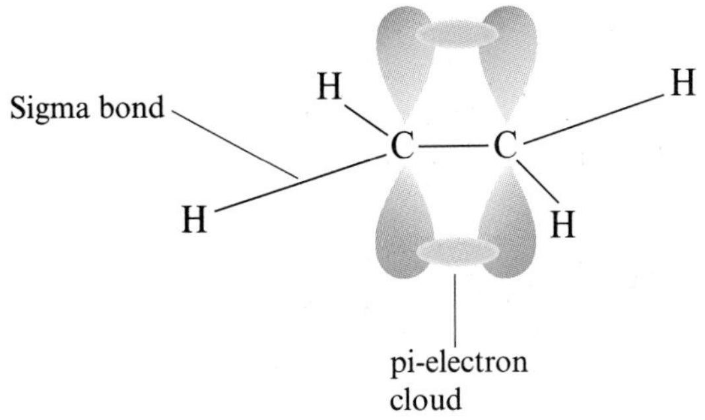

Figure 2.2.14
Sigma- and pi-bonds in ethene

The coverage of covalent bonding and overlap of orbitals is a study in its own right. To go further with this topic, you might like to read a more advanced text and consider the topic of 'hybridisation'.

Answers to diagnostic test

If you score less than 80%, then work through the text and re-test yourself at the end using this same test. If you still get a low score then re-work the unit at a later date.

1 Three atoms joined together by covalent bonds.   **(2)**

2 In each there is 1 point for: correct total number of atoms of each element, correct number of electrons shared by each atom.

a)

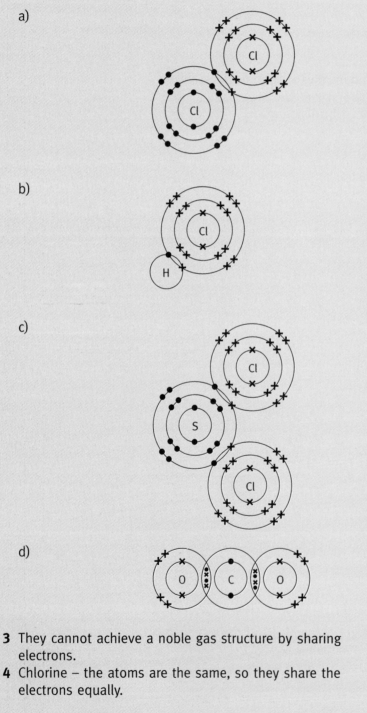

b)   **(3)**

c)   **(4)**

d)   **(5)**

                                                                        (5)

3 They cannot achieve a noble gas structure by sharing electrons.   **(1)**

4 Chlorine – the atoms are the same, so they share the electrons equally.   **(2)**

22 Marks   (80% = 18)

FURTHER QUESTIONS (NO ANSWERS GIVEN)

1 Show the outer electrons and bonding in the following compounds.

   a) Hydrogen and nitrogen in the compound hydrazine, NH_2NH_2.
   b) Hydrogen and oxygen in hydrogen peroxide, H_2O_2 (or HOOH).
   c) Carbon and hydrogen in methane, CH_4.
   d) Carbon and hydrogen in ethane, C_2H_6, or CH_3CH_3.

2 Compare the bonding in ethane with that in ethene.

Unit 2.3
Bonding and Properties

Starter check

What you need to know before you start the contents of this unit.

You should be familiar with the work in Module 1 and earlier units of Module 2.

Aims   **By the end of this unit you should understand that:**

- Bonds differ in strength.
- The type of bonding affects the properties of compounds.

Diagnostic test

Try this test at the start of the unit. If you score more than 80%, then use this unit as a revision for yourself and scan through the text. If you score less than 80% then work through the text and re-test yourself at the end by using this same test.

The answers are at the end of the unit.

1 Which would you expect to have the higher melting point, and why: magnesium oxide or sulfur dichloride? **(3)**
2 What is a giant structure? **(2)**
3 Which is stronger, a covalent C–C bond or forces between two adjacent covalent molecules? **(2)**
4 Which is stronger, the attraction between ions or the attraction between covalent molecules? **(2)**
5 Which of the following would you expect to conduct electricity?

a) Solid carbon dioxide;
b) liquid carbon tetrachloride;
c) solid sodium chloride;
d) liquid potassium hydroxide;
e) sulfur dioxide gas. **(5)**
6 Which of the materials in Question 5 would you expect to be soluble in water? **(5)**

19 Marks   (80% = 15)

2.3.1 Lattices

When we consider the types of bonding which occur between metals and other atoms, then ionic bonding is the most probable type of bonding. See also Unit 2.1.

When sodium reacts with chlorine, they combine ionically to form sodium chloride, NaCl, which is made up of sodium ions, Na^+ and chloride ions, Cl^-. In any real sample of sodium and chlorine there will be billions of atoms producing billions of positive and negative ions. Opposite charges attract, so each sodium ion will try to attract as many chloride ions around itself as it can, and repel other sodium ions. The chloride ions will be trying to do the opposite. All this attraction and repulsion ends up with the ions arranged in a regular, cubic arrangement in which each sodium ion has six chloride ions as its nearest neighbours, and each chloride ion has six sodium ions as its nearest neighbours, as shown in Figure 2.3.1.

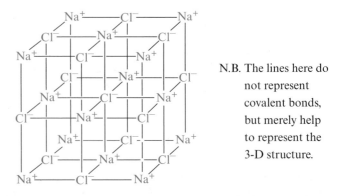

N.B. The lines here do not represent covalent bonds, but merely help to represent the 3-D structure.

Figure 2.3.1
The sodium chloride crystal lattice

The actual arrangement will depend on the relative sizes and charges on the ions involved, but there will always be a regular arrangement, called a **crystal lattice**. This is generally true for any ionic compound, except that the type of arrangement and overall shape will be different in different materials.

In the lattice of a sodium chloride crystal, there is no individual particle you can pick on and say, 'this is a particle of sodium chloride.' The lattice is what is known as a **giant structure** – it is made up of billions of ions arranged in a regular order and held together by electrostatic attraction of the positive and negative ions. Note that there is an equal number of repulsions between ions which have the same charge, but the oppositely-charged ions are closer together, so there is overall attraction which holds the lattice together.

2.3.2 Giant Covalent Structures

You get giant structures that are held together by covalent bonds as well. Think about carbon. Its electronic configuration is $1s^2 2s^2 2p^2$; it has four electrons in its outermost energy level. Each carbon atom would require to have four more electrons to give it an inert gas structure. They can achieve this by forming covalent bonds (Figure 2.3.2). The middle carbon has a share in an inert gas structure, but the other four do not. We need to add some more carbon atoms (Figure 2.3.3). You can go on like this in two dimensions, building up a giant structure, held together by covalent bonds.

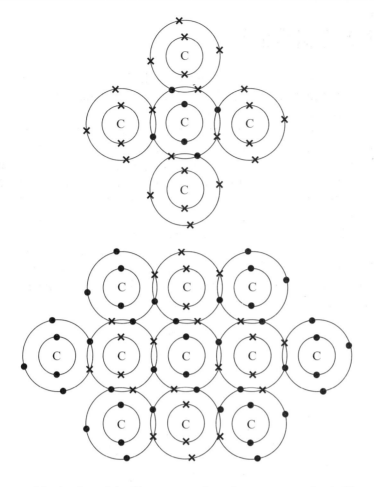

Figure 2.3.2
Carbon forms covalent bonds

Figure 2.3.3
The impossible structure of
carbon

The trouble is that this diagram makes the structure look like a flat layer. The structure shown in Figure 2.3.3 is quite wrong. In fact, the four covalent bonds of carbon are arranged tetrahedrally around the carbon atom so that the actual structure is as shown by Figure 2.3.4.

This is the basis of the structure of diamond, which is made up of carbon atoms in tetrahedral units covalently bonded together into what is, in effect, a huge molecule.

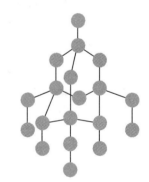

Figure 2.3.4
Tetrahedral arrangement of
carbon atoms in diamond

SUMMARY

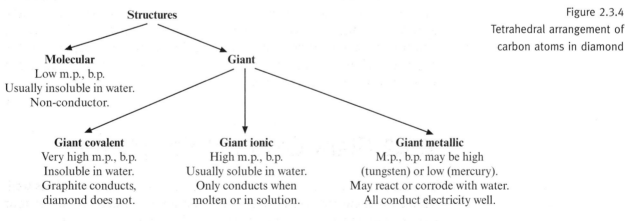

Structures

Molecular
Low m.p., b.p.
Usually insoluble in water.
Non-conductor.

Giant

Giant covalent
Very high m.p., b.p.
Insoluble in water.
Graphite conducts,
diamond does not.

Giant ionic
High m.p., b.p.
Usually soluble in water.
Only conducts when
molten or in solution.

Giant metallic
M.p., b.p. may be high
(tungsten) or low (mercury).
May react or corrode with water.
All conduct electricity well.

2.3.3 Giant Metallic Structures

The two main physical properties of metals are that they conduct electricity without being broken up (at most, they may get hot – hence electric fires, kettles, and tungsten filament light bulbs), and they can be shaped – hammered, pressed, bent, drawn into wires – without breaking.

These two properties make common metals like iron (in steel) and copper into some of the most useful materials which underpin civilisation. Without steel, there would be virtually no transport, no large bridges, no high buildings. Without copper, no electrical wiring and little electrical machinery.

These two properties reflect the essential structure of metals. They consist of a regular lattice (see Section 2.3.1) in which **the metal atoms have lost their outer electrons**. So the metal **ions** are arranged in regular layers among a 'sea' of loose electrons. If extra electrons are put into one end of a metal wire, instantaneous repulsions mean that an equal number leave the other end. Hence electrical conduction. If the metal is put under stress, a layer of positively charged metal ions can slide over the layer underneath – but **both layers are still attracted to the negatively-charged electrons between them**. So metals can bend and be shaped without breaking.

Sodium, magnesium and aluminium have one, two and three electrons respectively in their outer shells. Can you see why sodium is the softest and aluminium the hardest of these three, and why aluminium is the best conductor?

2.3.4 Different Bonding, Different Properties

MELTING AND BOILING POINTS

How do the differences in properties between diamond and sodium chloride reflect their bonding? First let us look at their melting points. Diamond, on heating to 4198 K, changes straight to a gas, whereas sodium chloride melts at 1074 K, a much lower temperature. What does this tell us about the strength of the forces holding together the particles in the two lattices? Since solids melt when their lattices break down, and since it takes a lot more heat energy to break down the diamond lattice, it follows that the forces holding together the diamond lattice, *i.e.* atom-to-atom covalent bonds, must be very strong. The electrostatic forces holding together the ions in an ionic lattice are held by opposite charge attraction and also can be very strong. But covalent and ionic bonds are different, and it is very difficult to compare 'like with like' because a compound is either covalent or ionic. Now let us think about tetra-chloromethane (carbon tetrachloride). It is also a covalent compound (Figure 2.3.5).

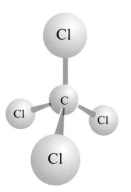

Figure 2.3.5
Tetrachloromethane

We know that the covalent bonds holding together the carbon and chlorine atoms **within** the molecules are very strong. But there are only small attractive forces **between** one molecule of carbon tetrachloride and the next one. Tetrachloromethane, CCl_4, is a liquid as the attractive forces between the molecules are low. In ionic compounds the particle-to-particle attraction is very strong.

This compound is a liquid at room temperature, that is to say it has a much lower melting point than sodium chloride. It follows that the forces holding together the sodium chloride lattice must be stronger than the forces holding together the particles of the solid in a carbon tetrachloride lattice. The particles of the carbon tetrachloride are individual molecules of CCl_4; the particles of the sodium chloride are charged ions all surrounded by thousands of others in one big lattice. It follows, then, that the forces between ions are stronger than the forces **between** non-charged covalent molecules. It also follows from this that covalent materials will, in general, have lower melting and boiling points than ionic compounds (unless, that is, the covalent materials have a giant structure so large that it is impossible to break down its structure by heat, *e.g.* diamond). Covalent compounds are usually gases or liquids or low melting or boiling solids. Ionic compounds are usually high melting solids. Giant covalent structures are always solids which cannot be easily melted or boiled, *e.g.* diamond.

ELECTRICAL CONDUCTIVITY

Another difference in properties between ionic and covalent compounds is electrical conductivity. No solid compound will conduct electricity, because the particles are not free to move. They are locked into the lattice.

If you melt the compound, or dissolve it in water, the particles become free to move. You can dip a positive and a negative electrode into the melt or the solution and see if it conducts. What you find is that covalent compounds do not conduct electricity, whereas ionic ones do. Once again this result is not unexpected, since covalent compounds are composed of neutral molecules which will not be attracted to the electrodes, whereas ionic compounds are made up of charged ions which **will** be attracted to them.

Conduction in metals causes no chemical change, because only electrons move (see Section 2.3.3).

Summary

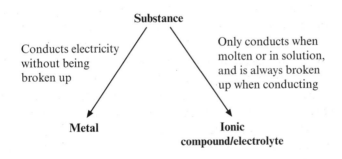

SOLUBILITY IN WATER

The third property to consider is solubility in water. On the whole, ionic compounds tend to be soluble in water, whereas covalent ones do not. This is because water molecules are not themselves absolutely neutral, but have a slight negative charge on the oxygen atom and slight positive charges on the hydrogen atoms (remember Section 2.2.3 on page 61). Thus, when an ionic compound is placed in water, the water molecules are attracted to the ions on the outside of the lattice and, together with a little stirring and/or heating, help to pull them off the lattice and into solution. Covalent compounds, on the other hand, are made up of neutral molecules, or molecules with only slight charges, and so are not as strongly attractive to water molecules. Thus, on the whole, ionic compounds tend to be much more soluble in water than covalent ones.

You should note that all of these property differences are generalisations. Although ionic compounds tend to have higher melting points than covalent ones, you do get some ionic compounds with low melting points and *vice versa*. In fact, water, which is covalent, has a much higher melting and boiling point than you might expect simply from the size of its molecules; hydrogen sulfide, for example, which has molecules almost twice as massive, is a gas at room temperature. The reason is that the separation of charge within the water molecule is much greater than in the hydrogen sulfide molecule, so the attraction between water molecules is also greater (so much greater that it is sometimes given the name 'hydrogen bonding').

You might find it helpful to make a table of the similarities and differences between covalent and ionic compounds.

2.3.5 Hydrogen Bonding

This term is given to a special type of attraction, usually only shown between hydrogen and either oxygen, nitrogen or fluorine. It is not like the usual covalent bonding, but is a form of attraction between small positive and negative charges in some polar molecules. It is a particularly important form of bonding as it occurs in the most important compound of Earth, water, and in the compound which makes each one of us unique, DNA.

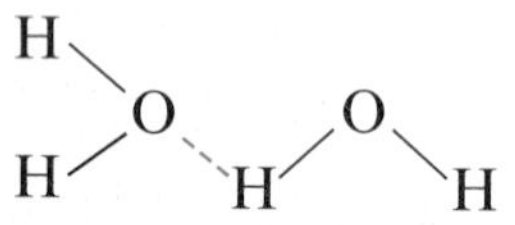

Water is made up of covalent molecules, H_2O (Figure 2.3.6). Its structure shows that there are sigma covalent bonds between the O and the H, and the angle between the O–H bonds is rather more than a right angle. Because oxygen is way over to the right-hand side of the periodic table, it tends to draw the electrons of the H–O bond towards itself. This causes the O in the molecule to be slightly more electron rich (and hence slightly more negative) than the deprived H which is left slightly poorer in its share of electrons (and hence slightly positive). This all occurs within a single molecule, but the neighbouring molecule will be the same, and so between the two molecules there will be a mutual attraction between the positive hydrogens and the negative oxygens. This attractive force in water usually means that 3 or 4 molecules of water group together. The energy needed to break these hydrogen bonds is much less than that required to break a conventional O–H covalent bond, but energy is still needed. In the case of water changing to steam, an indication of the energy needed is that the boiling point of water is 373 K, whereas the hydrides of other light first-period elements have much lower boiling points. Similarly, the much heavier hydrogen sulfide is still gaseous at room temperatures as it cannot form these hydrogen bonds.

Much more could be said about the importance of hydrogen bonding, as it is responsible for the low density of ice and for keeping proteins in shape in the cells of our bodies. A reference to hydrogen bonding occurs in the study of alcohols in Unit 7.1.

Figure 2.3.6
Hydrogen bonding in water

Answers to diagnostic test

If you score less than 80%, then work through the text and re-test yourself at the end using this same test. If you still get a low score then re-work the unit at a later date.

1 Magnesium oxide – it is ionic – the forces holding the solid particles together are stronger due to the particles being charged ions, whereas sulfur chloride is covalent and there are only weaker intermolecular forces between the particles. **(3)**

2 A giant structure is one in which there is an indefinitely repeating group of chemically-bonded particles, usually in three dimensions. Examples are diamond, graphite, any crystalline metal, and any solid ionic compound. **(2)**

3 Covalent, C–C bonds. **(2)**

4 Stronger between ions. **(2)**

5 The only conductor will be liquid potassium hydroxide. (5 marks – knock off 1 point for any others you put down that are incorrect). **(5)**

6 a) No
   b) No
   c) Yes
   d) Yes
   e) No **(5)**

19 Marks (80% = 15)

FURTHER QUESTIONS (NO ANSWERS GIVEN – FURTHER AND MORE ADVANCED READING MIGHT BE REQUIRED TO ANSWER THESE QUESTIONS)

1 Would you expect potassium chloride to have properties and a crystal lattice similar to those of sodium chloride?

2 Explain the plate-like covalent structure of the form of carbon called graphite.

3 A ball-like structure of carbon has recently been discovered called Buckminsterfullerene. It has 60 carbon atoms arranged like the patches on the surface of a football. Would you expect this to be covalent or ionic? Would you call this a giant structure?

4 Calcium chloride is ionic and forms a crystal lattice, but could it be arranged like the structure of sodium chloride? What alternatives are there? There are two chloride ions to every one calcium ion.

5 Copper sulfate is a beautiful crystalline structure of formula $CuSO_4 \cdot 5H_2O$ The ratio of copper to sulfate is 1:1, so why does it not have a cubic crystal lattice like NaCl?

Unit 2.4
Electronic Configuration and Reactivity

Starter check

What you need to know before you start the contents of this unit.

You will need to understand the principles of bonding as set out in Units 2.1 and 2.2.

Aims **By the end of this unit you should understand that:**

- Some elements react more readily than others.
- Their electronic configuration affects their reactivity.

Diagnostic test

Try this test at the start of the unit. If you score more than 80%, then use this unit as a revision for yourself and scan through the text. If you score less than 80%, work through the text and re-test yourself at the end by using this same test.

The answers are at the end of the unit.

1 Which is more reactive, potassium or magnesium, and why? (7)
2 Which is more reactive, sulfur or fluorine, and why? (7)
3 Why might you expect sulfur and bromine to have similar reactivities? (7)
4 Why can aluminium not be extracted from its oxide ore using coke as a reducing agent? (3)
5 Why does copper not react with dilute hydrochloric acid? (3)

27 Marks (80% = 22)

2.4.1 Loss of Electrons

We have seen that atoms like to get a noble gas configuration when they react so that the compound formed will be stable (Units 2.1 and 2.2). Some atoms can achieve this by giving away or losing electrons. What holds the electrons in the atom in the first place?

The nucleus is positively charged; electrons are negatively charged, so they are attracted by the nucleus, and this attraction holds them in their 'orbit'. If an electron can get some extra energy from somewhere, it can jump to an orbit further away from the nucleus. If it gets sufficient energy, it may be able to escape from the atom altogether. It is a bit like a rocket leaving Earth: the rocket has to fight against the force of gravity which is pulling it back down. If there is enough energy produced in burning the rocket fuel, it will be able to escape; if not, it will fall back to Earth.

The force of gravity gets less the further you are away from Earth. It is the same with electrostatic attraction between the electrons and the nucleus. The further away an electron is from the nucleus, the less energy it takes to escape from the atom. The bigger an atom is, then, the further away from the nucleus its outermost electrons are; in addition, the electrons in the innermost orbits 'shield' the outermost electrons from the attractive forces of the positive charge on the nucleus, thus for atoms which have the same number of electrons in the outermost energy level, the bigger an atom is the easier it is for the atom to get a noble gas structure by losing electrons. They are held more loosely.

In Group 1 of the periodic table, the first element is lithium. Its electronic configuration is 2:1 (two electrons in the first shell and one in the second). To get a noble gas structure, it has to lose the electron from the outermost shell. The next element is sodium, with configuration 2:8:1. Sodium also has to lose one electron, but this electron is in the third shell from the nucleus, and thus easier to lose. The next element, potassium, with configuration 2:8:8:1, again has to lose one electron, but this electron is in the fourth shell, still further away from the nucleus and thus even easier to lose.

In other words, as you go down a group in the periodic table, it gets easier and easier for atoms to lose an electron and form a positive ion. This is true for any group, not just for Group 1.

Let us have a look at the comparison between Group 1 and Group 2. Sodium (Group 1), has the configuration 2:8:1. Magnesium (Group 2), has the configuration 2:8:2. Sodium atoms have to lose one electron to get a noble gas configuration. Magnesium atoms have to lose two electrons to get a noble gas configuration. It is fairly obvious that it is going to take more energy for two electrons to escape than for one. You might think at first that it will take twice as much energy; in fact it takes more than twice as much. Think of it this way: the electrons to be lost are in the third shell, so they ought to be the same distance from the nucleus as the electrons in the sodium atom, but this isn't true. The magnesium nucleus (atomic number 12) has got more protons than the sodium nucleus (atomic number 11), and so all the electrons in a magnesium atom are pulled slightly closer to the nucleus than the electrons in a sodium atom. Being slightly closer, it's somewhat harder for them to escape. But there is another thing. The first electron to escape from a magnesium atom is escaping from a neutral atom; it is a bit harder for it to get out than for an electron to leave a sodium atom, but not that much harder. When the first electron has gone, the magnesium atom is no longer neutral – it has become a positively charged ion; so the second electron is trying to escape from something that is already positively charged, and that's a lot harder than to escape than from a neutral atom. It takes quite a lot more energy for one magnesium atom to lose two electrons than it does for two sodium atoms to lose one each; more than four times as much.

In general, then, as you go across a row, or period, in the periodic table, it gets harder for atoms to lose electrons and become positive ions. An indication of this is the energy needed to take away the first electron from the neutral atoms. The property is called the first ionisation energy, and a graph of first ionisation energy against atomic can be constructed from the data given in Table 2.4.1.

Table 2.4.1 *First ionisation energy and atomic radius data for the first twenty elements*

Name	Atomic number	First ionisation energy/kJ mol^{-1}	Atomic radius/pm
Hydrogen	1	1310	32
Helium	2	2372	50
Lithium	3	520	152
Beryllium	4	899	112
Boron	5	801	98
Carbon	6	1086	91
Nitrogen	7	1402	92
Oxygen	8	1314	73
Fluorine	9	1681	72
Neon	10	2081	70
Sodium	11	496	186
Magnesium	12	738	160
Aluminium	13	577	143
Silicon	14	787	132
Phosphorus	15	1012	128
Sulfur	16	1000	127
Chlorine	17	1251	99
Argon	18	1520	98
Potassium	19	419	227
Calcium	20	590	197

All ionisation energies have *positive values – i.e.* they are **endothermic**. Energy is required to pull an electron off *any* atom.

2.4.2 Gain of Electrons

Elements in Group 7 of the periodic table have seven electrons in their outermost shells. From what you have read in Section 2.4.1 above you will see that it is virtually impossible for them to get a noble gas structure by losing seven electrons; it is much easier for them to gain one electron.

On the face of it, in fact, it should take no effort at all for any atom to take in an electron. After all, the nucleus is positively charged, so you might expect it to attract any electrons it comes across.

Actually, the picture isn't quite as simple as that (it never is, is it?). Think about bromine. Its configuration is 2:8:18:7. Imagine an extra electron entering a bromine atom. It is attracted towards the nucleus and starts to move towards it; but between the incoming electron and the nucleus there are energy levels containing 35 electrons, each of them negatively charged and therefore repelling the newcomer.

Now think about chlorine, configuration 2:8:7. An incoming electron is faced by repulsion from only 17 electrons – a lot less than for bromine. In addition, the newcomer can get closer to the positive nucleus, since it is entering the third shell rather than the fourth. Thus, chlorine atoms find it easier to take in an electron and form a negative ion than do bromine

Putting the first electron into a neutral atom is **exothermic** – energy is given out, as the attraction of the positive nucleus can still be felt at the surface of the atom.

atoms. In other words, as you go up any group in the periodic table and the atoms get smaller, it becomes easier to form a negative ion. This is shown by plotting the values for the atomic radii given in Table 2.4.1.

Let us compare Groups 6 and 7. Oxygen (Group 6) has configuration 2:6; fluorine (Group 7) has configuration 2:7. Oxygen has to gain two electrons and fluorine only one in order to get a noble gas configuration. Again, as with loss of electrons, it is fairly obvious that it will take more energy to force two electrons into an atom than to insert only one. The first electron to enter an oxygen atom changes the atom to a negative ion. The second electron is thus trying to enter something that is already negatively charged and therefore much harder for a negative particle to get into. This means that as you go across a row, or period, from right to left, it becomes progressively harder for the elements to form negative ions in which the outer shell is full of electrons – *e.g.* F^- is easier to form than O^{-2} or N^{3-}.

2.4.3 Trends in the Periodic Table

Elements in the groups on the left-hand side of the periodic table tend to react by losing electrons to form positive ions. As you go down a group, it gets easier for atoms to lose electrons. In other words, as you go down a group of metals it gets easier for the elements to react; they become more reactive (see Figure 2.4.1). Another way of saying this is that **reactivity increases as you go down a group on the left-hand side of the table**.

Similarly, as you go across a period from left to right, atoms have to lose more electrons to form stable positive ions. The more electrons have to be lost, the more energy it takes, so the harder it gets to form positive ions. In other words, as you go **from left to right across a period on the left-hand side of the table, reactivity decreases**.

In the same way, on the right-hand side of the table elements tend to react by gaining electrons. As you go down a group, it gets harder to gain electrons and reactivity decreases. Similarly, as you go from right to left across a period, more electrons have to be gained which makes it harder, so reactivity decreases. **So, on the right-hand side of the table, reactivity increases up the group, and decreases from right to left in a period**.

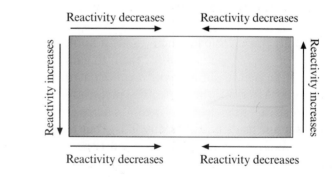

Figure 2.4.1
General reactivity trends
in the periodic table

2.4.4 Reactivity Series

If you look at Group 2 in the periodic table, you will see that calcium falls below magnesium; calcium is therefore more reactive than magnesium. Sodium falls to the left of magnesium, in the same period; sodium is therefore more reactive than magnesium. You might expect, therefore, that calcium and sodium will be fairly alike in terms of reactivity. This is, indeed, the case. They both, for example, react quite vigorously with cold water, corrode rapidly in air and burn when heated in air. Sodium, however, reacts more vigorously than calcium. We can therefore arrange these three elements in order of reactivity thus:

Sodium (most reactive)
Calcium
Magnesium (least reactive)

Thus, by looking at the positions of the elements in the periodic table and their reactions we can build up a **reactivity series**, with the most reactive elements at the top and the least reactive at the bottom. We can include as many or as few elements as we wish in this series. If we only include the most common elements, we arrive at a list which looks like this:

Potassium (K)
Sodium (Na)
Calcium (Ca)
Magnesium (Mg)
Aluminium (Al)
Carbon (C)
Zinc (Zn)
Iron (Fe)
Lead (Pb)
Hydrogen (H)
Copper (Cu)
Silver (Ag)
Gold (Au)

Such a series is very useful in explaining and predicting the outcomes of various reactions.

One example is the **displacement reaction**. Aluminium is more reactive than iron; to put it another way, aluminium forms positive ions more easily than iron does. Iron(III) oxide is an ionic compound: it contains positive iron ions and negative oxygen ions. If you mix aluminium metal with iron oxide, and provide some energy by heating the mixture, the aluminium atoms will give up electrons and force them on to the iron ions, thus converting the iron oxide into metallic iron:

$$2Al + 2Fe^{3+} \rightarrow 2Al^{3+} + 2Fe$$

To put it another way:

$$2Al + Fe_2O_3 \rightarrow Al_2O_3 + 2Fe$$

In other words, a more reactive element will displace a less reactive one from its compounds. If, for example, you dip a copper wire into silver nitrate solution and leave it for a while, crystals of silver will start to grow on the copper wire. The copper is more reactive than the silver and forces electrons on to the silver ions, thus converting them into metallic silver atoms:

$$Cu + 2AgNO_3 \rightarrow Cu(NO_3)_2 + 2Ag$$

or

$$Cu + 2Ag^+ \rightarrow Cu^{2+} + 2Ag$$

This reaction of aluminium with iron(III) oxide is called the **thermite reaction**. It is used to make molten steel for mending and welding broken railway lines; the temperature reached is about 2000 °C. The thermite mixture was often used in World War II incendiary bombs, such as those which devastated London, Hamburg and Dresden.

In the same way, any metal below carbon in the reactivity series can be extracted from its ore by heating the ore with carbon, because the carbon will displace the metal. The most important of these reactions, because our civilisation depends so much on steel, which is an alloy of iron, is the production of iron by heating iron ore (iron oxide) with coke (which is a form of carbon). The carbon, being more reactive than iron, displaces it thus:

$$3C + Fe_2O_3 \rightarrow 3CO + 2Fe$$

Again, you can tell that copper, silver and gold will not be attacked by *dilute* acids. This is because they are all less reactive than hydrogen and will thus not displace hydrogen from an acid. On the other hand, any metal above hydrogen in the reactivity series will react with a dilute acid. If, for example, you add magnesium to dilute hydrochloric acid, the magnesium will displace the hydrogen from the acid as follows:

$$Mg + 2HCl \rightarrow MgCl_2 + H_2$$

or

$$Mg + 2H^+ \rightarrow Mg^{2+} + H_2$$

You can see how important the reactivity series is in predicting the outcome of displacement reactions.

There are other uses as well; *e.g.* some compounds, when heated, will decompose. This is called **thermal decomposition**. Mercury oxide is a good example; it is made up of positive mercury ions and negative oxygen ions (Hg^{2+} and O^{2-}). Mercury is very unreactive: it comes between copper and silver in the reactivity series. It was reluctant to give up its electrons in the first place to form positive ions. If you heat mercury oxide, the oxygen ions give back electrons to the mercury ions. The mercury ends up as metallic mercury, and the oxygen ions end up as neutral, gaseous oxygen molecules:

$$2HgO \rightarrow 2Hg + O_2$$

or

$$2Hg^{2+} + 2O^{2-} \rightarrow 2Hg + O_2$$

On the other hand, if you heat sodium oxide, nothing happens. This is because sodium is towards the very top of the reactivity series. It was eager to form positive ions in the first place by giving away electrons, so sodium ions are less likely than mercury ions to accept the electrons back again and form neutral atoms. The general rule is that the more reactive a metal is, the more stable its compounds are likely to be, *i.e.* the less likely to be split up by heating.

Bronze contains copper and tin. Why do you think the Bronze Age came before the Iron Age?

Because the reactivity series is so important as a tool for thinking about the outcome of reactions, it is worthwhile trying to remember it. The following is a mnemonic for the series, using only the elements in the series printed out above, *i.e.* if you remember this sentence, each word starts with the same letter as the corresponding element. You can either use this mnemonic or make up your own: you can use an extended reactivity series, putting in other elements. The general rule for mnemonics is that the more ridiculous the picture, the better they are likely to be at reminding you of the order.

Perhaps	Potassium
Some	Sodium
Cows	Calcium
May	Magnesium
All	Aluminium
Come	Carbon
Zooming	Zinc
In	Iron
Large	Lead
Herds	Hydrogen
Chewing	Copper
Some	Silver
Grass	Gold

You will notice that almost all the elements in this series are metals. The same sorts of arguments work for the non-metals. The halogens provide a good example. The order of reactivity for these is the same as the order of the elements in the group:

Fluorine
Chlorine
Bromine
Iodine
Astatine

Thus, for example, if you pass chlorine gas through potassium bromide solution, the solution will go reddish-brown because the chlorine displaces the bromine from the potassium bromide:

$$2KBr + Cl_2 \rightarrow 2KCl + Br_2$$

or

$$2Br^- + Cl_2 \rightarrow 2Cl^- + Br_2$$

The higher element in the reactivity series displaces the lower one from a solution of its salt.

Answers to diagnostic test

If you score less than 80%, then work through the text and re-test yourself at the end using this same test. If you still get a low score then re-work the unit at a later date.

1 Potassium is more reactive as it only has to lose one electron – Mg has to lose two. It is more difficult to lose two than to lose one – K's outermost electron is further from the nucleus than Mg's is – less attraction to nucleus – easier to lose.   (7)

2 Fluorine is more reactive as it only has to gain one electron – S has to gain two. It is more difficult to gain two than to gain one – F's outermost electron is closer to the nucleus than S's is – more attraction to nucleus – easier to gain.   (7)

3 S is the third period – Br is the fourth – as you go from the third to the fourth period on right-hand side of the table, reactivity decreases. Br is on the extreme RHS – S is further to the left – as you go from left to right on the RHS of the table, reactivity increases. Thus, reactivities are roughly the same.   (7)

4 Coke is carbon. Aluminium is more reactive than carbon, so carbon cannot displace aluminium from its ore.   (3)

5 All acids contain hydrogen. Copper is less reactive than hydrogen, so copper cannot displace hydrogen from dilute acids.   (3)

27 Marks   (80% = 22)

FURTHER QUESTIONS (NO ANSWERS GIVEN)

1 Why are copper, silver and gold so widely used in coinage and jewellery?

2 Why is aluminium used for so many things if it is so reactive towards oxygen?

3 Why are the metals copper and lead used for some forms of roofing?

4 Why is zinc used to 'galvanise' iron sheets to protect them from the weather?

Unit 2.5
Oxidation Numbers

Starter check

What you need to know before you start the contents of this unit.

You will need to have a good understanding of chemical reactions as given in Units 2.1–2.4.

Aims **By the end of this unit you should know and understand that:**

- Elements combine in ratios that reflect their electronic structures.
- The formula of a compound tells you the ratio of atoms of the different elements of which it is composed.
- Elements can be assigned oxidation numbers which tell you their 'combining power'.
- Oxidation numbers can be used to write formulae.

Diagnostic test

Try this test at the start of the unit. If you score more than 80%, then use this unit as a revision for yourself and scan through the text. If you score less than 80% then work through the text and re-test yourself at the end by using this same test.

The answers are at the end of the unit.

1 What is the oxidation number of the following elements?

a) Cl^-
b) Cl_2
c) Na^+
d) Fe^{3+}
e) the sulfur in the sulfate ion, whose formula is SO_4^{2-}
f) manganese in MnO_4^- **(6)**

2 Why do the transition metals have variable oxidation numbers? **(5)**

3 Write formulae for the following compounds:

a) potassium oxide
b) calcium nitride(III)
c) iron(II) chloride
d) aluminium sulfate
e) scandium(III) nitrate **(15)**

26 Marks   (80% = 21)

2.5.1 Oxidation Numbers and Group

Elements combine to form compounds. Oxygen comprises roughly 20% of the atmosphere. Most metals, if left exposed to the air, will tarnish or corrode by combining with oxygen. The more reactive a metal is, the faster it will corrode (this is why copper, silver and gold, down at the bottom of the reactivity series are so widely used as coinage and jewellery metals – being so unreactive they only corrode very slowly). Many foodstuffs go off or become 'rancid' because they combine with oxygen from the air. This combination with oxygen, because it is such a common and important reaction, is given a special name. It is called **oxidation**.

Think about what is actually happening when oxygen combines with a metal. Oxygen, being in Group 6, has six electrons in its outer shell. When it reacts, it takes in two electrons to get a noble gas structure, thus forming a negative ion, O^{2-}. The metal with which it reacts must therefore lose electrons to form a positive ion. Oxidation can thus be defined as gain of oxygen or loss of electrons.

REMEMBER: OIL 'Oxidation Is Loss' of electrons. The opposite of oxidation is **reduction**, so remember OILRIG, 'Oxidation Is Loss, Reduction Is Gain' of electrons. Oxidation and reduction *always* occur together; such reactions are known as **REDOX** reactions.

Sodium forms positive ions with one positive charge. So do all the other elements in Group 1. So a sodium atom going to the sodium ion is a loss of one electron; it has been oxidised by 1 (by electron loss). This can be summed up by saying that the Group 1 elements in their compounds have oxidation number +1. Similarly, the Group 2 elements in their compounds have oxidation number +2, in Group 3, +3.

The atoms of elements in Group 7 have seven electrons in their outermost energy level and so react by taking in one electron to form ions with one negative charge. They have gained electrons, and are so reduced (the opposite to oxidation). This can be summed up by saying that the Group 7 elements in their compounds have oxidation number −1. Similarly, the elements in Group 6 have oxidation number −2. In fact, there is an obvious pattern in the oxidation numbers of the elements, which goes like this:

Group Number	1	2	3	4	5	6	7
Oxidation Number	+1	+2	+3	+4	−3	−2	−1

The term 'oxidation number' has been used in recent years to explain more fully what goes on in redox reactions, particularly in reactions that do not use oxygen at all but where there is a movement of electrons.

N.B. The oxidation number of **all** elements which are not combined with anything else is **zero**. So the oxidation number of both sodium metal and chlorine gas is zero; but in sodium chloride, the oxidation number of sodium is +1 and of chlorine −1.

2.5.2 Oxidation Numbers and Formulae

You can use the link between group number and oxidation number to predict the formulae of compounds formed between the elements. Take sodium and chlorine, for example. Their oxidation numbers are $+1$ and -1, respectively, *i.e.* sodium forms ions with a single positive charge and chlorine forms ions with a single negative charge. Now the compound overall has got to be electrically neutral (after all, you don't get an electric shock from salt), so there must be just as many positive charges as there are negative ones. In other words, there must be one chloride ion for every one sodium ion, *i.e.* the formula must be NaCl. Another way of saying all of that would be that in sodium chloride (as in all other compounds) the sum of the oxidation numbers must be zero: $+1$ (for the sodium) $+ (-1)$ (for the chlorine) $= 0$.

Similarly, for magnesium chloride: magnesium's oxidation number is $+2$ and that of chlorine is -1, so $(+2) + [2 \times (-1)] = 0$, and there must be two chloride ions for every magnesium ion, so the formula must be $MgCl_2$.

In the same way, oxygen's oxidation number is -2, so when it combines with magnesium we have $+2 + (-2) = 0$, so there must be one oxygen for every magnesium, and the formula must be MgO.

Aluminium oxide is a bit tricky. The oxidation numbers are $+3$ and -2, respectively. How can we make those add up to zero? Well, $2 \times (+3) = +6$ and $3 \times (-2) = -6$, so there must be two aluminiums to every three oxygens and the formula is Al_2O_3. That is all right for ionic compounds, but what about covalent ones? It turns out that it doesn't make any difference whether the compound is ionic or covalent. Look at it this way: Group 7 elements can bond either ionically or covalently, but whichever they do they need one extra electron to give them a noble gas structure, so their oxidation number will still be -1. The same goes for the other groups. Carbon's usual oxidation number is $+4$, and oxygen's is -2, so $+4 + [2 \times (-2)] = 0$, and there must be two oxygens to one carbon, so the formula for the most common oxide of carbon is CO_2.

2.5.3 The Transition Metals

You've probably noticed that in everything we've said so far about electronic configuration, bonding and formulae we've missed out a big block of elements altogether. Those are the ones in the centre of the periodic table, starting with scandium in the top left and running to mercury in the bottom right. This block of elements is called the **transition metals**. What makes them special?

Argon has eight electrons in the third energy level. As we've seen, this is a rather special and stable arrangement. The third energy level can hold 18 electrons, so the next element (potassium) ought to have nine electrons in the third level. Instead, as if to delay spoiling the ideal noble gas structural arrangement, the extra electron goes into the fourth level, so potassium is 2:8:8:1.

Similarly, calcium is 2:8:8:2.

As if the strain has become too much, the next electron does go into the third level, which slowly fills up as you go from scandium across to zinc.

The ground state configurations (see Section 1.5.4) of the elements from scandium to zinc are shown below.

Scandium	2:8:9:2	$1s^2 2s^2 2p^6 3s^2 3p^6 3d^1 4s^2$
Titanium	2:8:10:2	$1s^2 2s^2 2p^6 3s^2 3p^6 3d^2 4s^2$
Vanadium	2:8:11:2	$1s^2 2s^2 2p^6 3s^2 3p^6 3d^3 4s^2$
Chromium	2:8:13:1	$1s^2 2s^2 2p^6 3s^2 3p^6 3d^5 4s^1$
Manganese	2:8:13:2	$1s^2 2s^2 2p^6 3s^2 3p^6 3d^5 4s^2$
Iron	2:8:14:2	$1s^2 2s^2 2p^6 3s^2 3p^6 3d^6 4s^2$
Cobalt	2:8:15:2	$1s^2 2s^2 2p^6 3s^2 3p^6 3d^7 4s^2$
Nickel	2:8:16:2	$1s^2 2s^2 2p^6 3s^2 3p^6 3d^8 4s^2$
Copper	2:8:18:1	$1s^2 2s^2 2p^6 3s^2 3p^6 3d^{10} 4s^1$
Zinc	2:8:18:2	$1s^2 2s^2 2p^6 3s^2 3p^6 3d^{10} 4s^2$

The slight differences for chromium and copper can be explained by the extra stability of a half-filled d sub-level (for chromium) and a full d sub-level (for copper).

Most of these elements have got two electrons in their outermost level, and you know that it is the electrons in the outermost level that determine properties such as oxidation number. In fact, these elements do have very similar properties, but one of those properties is that they don't have a fixed oxidation number. You might expect most of them to have an oxidation number of $+2$ because there are two electrons in their outer shells. In fact, these elements can very easily shift electrons between the third and fourth energy levels. If you look at copper, for example, you would expect it to have an oxidation number of $+1$, and it does – in some of its compounds. If a copper atom lets one of the electrons from its third level rise to the fourth level, however, and then loses both these electrons, it will form ions with a double positive charge, *i.e.* its oxidation number will be $+2$ in some of its compounds.

That means, for example, that there are two compounds called copper oxide, one of which has copper of oxidation number $+1$ and the other of oxidation number $+2$. How are we to know which is which? One way is (hopefully) by looking at the name. Their names are copper(I) oxide and copper(II) oxide. The Roman numerals in the names tell you what the oxidation number of copper in the compound is.

- Copper(I) oxide
  Copper $= +1$, oxygen $= -2$; $[2 \times (+1)] + (-2) = 0$, so there must be two coppers to each oxygen, so the formula is Cu_2O.

- Copper(II) oxide
  Copper $= +2$, oxygen $= -2$; $+2 + (-2) = 0$, so there must be one copper to each oxygen, and the formula is CuO.

The other way is by looking at them. All the transition metals tend to have coloured compounds, and the colour depends on the oxidation number of the metal. Copper(I) oxide is red and copper(II) oxide is black.

Unfortunately, there is no systematic way of working out the possible oxidation numbers of the transition metals, although it is quite easy to work out the *maximum* oxidation number. The colours on the other hand are a bit easier. Iron(II) compounds tend to be green and iron(III) compounds tend to be brown, for example – but even that is not universally true.

- Iron(II) oxide

  Iron = +2, oxygen = −2;   +2 + (−2) = 0, so there must be one iron to each oxygen, and the formula is FeO.

- Iron(III) oxide

  Iron = +3, oxygen = −2;   [2 × (+3)] + [3 × (−2)] = 0, so there must be two irons to every three oxygens, and the formula is Fe_2O_3.

This trick of promoting electrons within or between energy levels in the atom is, unfortunately, one that isn't limited to the transition metals. The elements in Groups 4 to 7 can do the same thing (Figure 2.5.1). Thus, there are two carbon oxides:

- Carbon(II) oxide (always known as carbon monoxide)

  Carbon = +2, oxygen = −2;   +2 + (−2) = 0, where there must be one carbon to each oxygen, so the formula is CO.

- Carbon(IV) oxide (always known as carbon dioxide)

  Carbon = +4, oxygen = −2;   +4 + [2 × (−2)] = 0, so there must be one carbon to two oxygens, and the formula is CO_2.

Similarly, there are two oxides of sulfur: sulfur(IV) oxide and sulfur(VI) oxide, known as sulfur dioxide and sulfur trioxide respectively.

- Sulfur(IV) oxide

  Sulfur = +4, oxygen = −2;   +4 + [2 × (−2)] = 0, so there must be one sulfur to two oxygens, and the formula is SO_2.

- Sulfur(VI) oxide

  Sulfur = +6, oxygen = −2;   +6 + [3 × (−2)] = 0, where there must be one sulfur to three oxygens, so the formula is SO_3.

Note that when oxidation numbers are included in a name, they are *always* represented by Roman numerals next to the name of the element involved. So, for example, $PbCl_2$ is lead(II) chloride; $KMnO_4$ is potassium manganate(VII). *Never* write oxidation numbers in the names of compounds of elements which cannot change their oxidation numbers, especially the metals in Groups 1 and 2. Thus, NaCl is just sodium chloride; $CaCO_3$ is calcium carbonate.

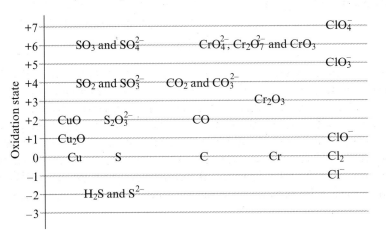

Figure 2.5.1

Typical oxidation states of common elements

2.5.4 Multi-atom Ions

All the compounds we've looked at so far have been binary compounds, *i.e.* they contain only two elements. Note that the names of binary compounds end in -ide. What if they contain more than two? Fortunately most non-binary compounds contain what are called multi-atom ions. These are groups of atoms that behave as if they were a single ion. Some examples are:

Name	Formula	Ox. state of whole ion
Sulfate	SO_4^{2-}	-2
Nitrate	NO_3^-	-1
Carbonate	CO_3^{2-}	-2
Hydroxide	OH^-	-1
Ammonium	NH_4^+	$+1$

To write the formulae of these compounds, all you have got to do is recognise the fact that they contain a multi-atom ion, look up its formula and oxidation number and follow the rules.

e.g. Sodium sulfate: sodium = $+1$, sulfate = -2.
$[2 \times (+1)] + (-2) = 0$, so there must be two sodiums to every sulfate, and the formula is Na_2SO_4.

What if there is more than one multi-atom ion present? Let us look at calcium nitrate.

e.g. Calcium = $+2$, nitrate = -1.
$+2 + [2 \times (-1)] = 0$, so there must be two nitrates to each calcium. We want to double the nitrate as a whole so, just as in maths, we put it in brackets like this: $Ca(NO_3)_2$. Similarly, ammonium sulfate is $(NH_4)_2SO_4$ and aluminium hydroxide is $Al(OH)_3$.

Just to complicate things a little further, there are actually two sulfate ions. That's because sulfur has different oxidation numbers. The sulfate ion given above is the sulfate(VI) ion, because the sulfur atom in it has an oxidation number of $+6$. The whole multi-atom ion has oxidation number -2, because there is one sulfur and four oxygens. Each oxygen has the oxidation number -2, the sulfur is $+6$, and so $+6 + [4 \times (-2)] = -2$.

There is another sulfate ion called the sulfate(IV) ion, because the sulfur in it has the oxidation number $+4$. Its formula is SO_3^{2-}, and it is usually still known as the *sulfite* ion. Each oxygen has an oxidation number of -2, so the oxidation number of the ion as a whole is $+4 + [3 \times (-2)] = -2$. So sodium sulfate(IV) is Na_2SO_3.

Similarly, there are nitrate(V) and nitrate(III) multi-atom ions. The former is the one we've come across, NO_3^-, with an oxidation number of -1, because $+5$ (for the nitrogen) $+ [3 \times (-2)] = -1$. This is usually simply called nitrate.

The nitrate(III) ion has a formula of NO_2^- and the nitrogen in it has an oxidation number of $+3$, so the ion has an overall oxidation number of $+3 + [2 \times (-2)] = -1$. So sodium nitrate(V) is $NaNO_3$ and sodium nitrate(III) is $NaNO_2$. The nitrate(III) ion is still usually known as the *nitrite* ion.

2.5.5 Summary of Oxidation and Oxidation Numbers

1 Hydrogen has an oxidation number (ON) of $+1$ and oxygen -2.

2 The ON of an element when not in a compound is 0 (zero).

3 The ON of an ion is the value and sign of the ion, *e.g.* $Na^+ = +1$, $Cl^- = -1$.

4 Some elements can have more than one ON depending upon the compound the element is in, *e.g.* S as an element is 0; S in H_2S is -2; S in SO_2 is $+4$; S in H_2SO_4 is 6.

5 ONs do not always have whole numbers, *e.g.* S in $Na_2S_4O_6$ is 2.5.

6 The overall ON of a compound is 0 (zero), *e.g.* water, $H_2O = [2 \times (+1)] + (-2) = 0$.

7 The overall ON of a multi-atom ion is the value of the charge, *e.g.* $SO_4^{2-} = -2$.

8 Oxidation is shown in a chemical reaction when the ON increases. Reduction is when the ON decreases.

You will be able to apply these rules to later reactions occurring in Module 3.

Answers to diagnostic test

If you score less than 80%, then work through the text and re-test yourself at the end using this same test. If you still get a low score, then re-work the unit at a later date.

1 a) -1
   b) 0
   c) $+1$
   d) $+3$
   e) $+6$
   f) $+7$
   (1 point each)   **(6)**

2 The penultimate energy levels are part-filled (2 points); electrons move easily between the two (1 point); the number of electrons in the outermost level changes (1 point); and hence the number of electrons lost to form ions changes (1 point).   **(5)**

3 a) K_2O
   b) Ca_3N_2
   c) $FeCl_2$
   d) $Al_2(SO_4)_3$
   e) $Sc(NO_3)_3$
   (3 points each)   **(15)**

26 Marks   (80% = 21)

FURTHER QUESTIONS (NO ANSWERS GIVEN)

1 What are the oxidation numbers of the following elements?

    a) Sulfur in copper sulfate.
    b) Magesium in magnesium chloride.
    c) Oxygen in hydrogen peroxide, H_2O_2.
    d) Nitrogen in the following compounds: ammonia, NH_3; hydrazine, N_2H_4; nitric acid, HNO_3.

2 Explain the oxidation changes occurring in the following chemical reactions.

    a) Oxygen reacting with magnesium to form magnesium oxide.
    b) Zinc reacting with dilute hydrochloric acid to form zinc chloride and hydrogen gas.
    c) Iron metal powder added to copper sulfate solution to form copper metal plus iron sulfate solution.

Module 3

Different Types of Chemical Reaction

It must be stated at the beginning of this module that chemical reactions can often be described in a number of different ways, and the descriptions used in this module are not the only ones. We shall see in Unit 3.1 that some reactions can be called 'combination reactions', but when the combination reaction involves oxygen then it can also be called 'oxidation' or 'reduction'. Other examples appear in each unit.

Unit 3.1
Synthesis and Combining

Starter check

What you need to know before you start the contents of this unit.

You should understand that chemical reactions occur when the outer electrons of elements inter-react. You will probably also have covered the ideas of elements in the periodic table. See particularly Module 2.

Aims **By the end of this unit you should understand that:**

- Elements combine to form compounds.
- When elements combine with oxygen, it is called oxidation.
- Combustion is a type of oxidation.
- Tarnishing of metals is a combination (and oxidation) reaction.
- Binary compounds have names ending in 'ide'.

Diagnostic test

Try this test at the start of the unit. If you score more than 80%, then use this unit as a revision for yourself and scan through the text. If you score less than 80%, then work through the text and re-test yourself at the end by using this same test.

The answers are at the end of the unit.

1 What is meant by the term 'binary compound'? What ending do the names of binary compounds all have? **(4)**

2 What will be the names of the compounds formed when the following pairs of elements combine?

   a) Magnesium and oxygen; **(2)**
   b) zinc and sulfur; **(2)**
   c) silicon and chlorine; **(2)**
   d) hydrogen and sulfur; **(2)**
   e) lead and bromine. **(2)**

3 When coal is burned, the products of combustion are carbon dioxide, water vapour and sulfur dioxide. Why? What does this tell you about the contents of the coal? How are these products formed? **(8)**

22 Marks (80% = 18)

3.1.1 Binary Compounds by Joining Elements Together

All the matter in the Universe is made up from just one hundred (approximately) elements. Everything you see around you is either elements, compounds or a mixture of these. All the compounds have been formed by these elements combining somehow over the years, some by direct contact and others by a series of devious routes.

Most of the elements are metals. The first tools that were made by human beings were made of stone, because that is all they had. Eventually metals were discovered. The first metal to be discovered was probably gold. Gold can be found naturally in rocks as the metal, un-combined with anything else (this is called being found 'native'). Most metals are found combined with other elements as ores, which are compounds.

There is a lot of oxygen in the atmosphere, and over the millions of years since the Earth was formed many metals have combined with oxygen to form compounds called oxides. Iron, for example, occurs in the molten core of our Earth, and is often found in the Earth's crust as the ore haematite, which is iron oxide. Iron oxide is a binary compound, which means it contains only two elements. Notice that its name ends with 'ide'. This is true for all binary compounds. Metals tend to combine with non-metals. Where a binary compound is made up of a metal and a non-metal, the first name of the compound is simply the name of the metal it contains and the second is the name of the non-metal, changed to end in 'ide'; for example, the compound between copper and chlorine is copper chloride.

In volcanic regions, natural un-combined sulfur is often found. Many metals have combined with this sulfur to form compounds called sulfides. Lead, for example, is often found as the ore galena, which is lead sulfide. The percentages of lead and sulfur in galena are always the same.

Bronze contains two elements, but it is not a compound. The elements are both metals, namely copper and tin, but they are not chemically combined in bronze. Bronze is a mixture of metals, an **alloy**; the two metals can be separated. The percentages of copper and tin in bronze can vary considerably, depending on how much of each you add together. The properties of bronze depend on the relative amounts of the two metals. Another very common alloy is brass (copper and zinc). In compounds, the ratio of the elements present is always the same value and fixed.

If a piece of zinc is left exposed to the air, it will eventually go dull, or tarnish. This is because the surface of the zinc becomes coated with a layer of zinc oxide. The same is true of most (but not all) metals. When an element combines with oxygen, we say it has been **oxidised**, or that an oxidation reaction has taken place. This reaction can be described as combination or alternatively as a redox reaction. More details of oxidation and oxidation states are listed in Unit 2.5.

$$2Zn(s) + O_2(g) \rightarrow 2ZnO(s)$$

$$C(s) + O_2(g) \rightarrow CO_2(g)$$

Whereas two metals are unlikely to combine chemically with each other, two non-metals may. Sulfur and oxygen, for example, combine to form sulfur dioxide. Carbon and oxygen combine to form carbon dioxide.

When two or more elements directly combine together to form a single new compound, this is usually called **synthesis** of the new compound. The term synthesis can also apply to much more complicated reactions involving a number of elements.

3.1.2 Fuels Combining with Oxygen

Most of the natural fuels we use are what are called 'fossil fuels' – they are the remains of organisms which were alive millions of years ago. All living things contain carbon. When anything burns, the elements in it combine with oxygen from the air. When these fuels burn, the carbon in them is oxidised to carbon dioxide. They all contain hydrogen as well, and when they burn the hydrogen is oxidised to water. For example, when methane (in natural gas) is burned, the equation for the reaction is:

$$CH_4(g) + 2O_2(g) \rightarrow CO_2(g) + 2H_2O(g)$$

3.1.3 Organic Synthesis

In Organic Chemistry (the chemistry of carbon compounds other than carbonates and the oxides of carbon), as you will see later in Modules 6 and 7, the term 'synthesis' often means something wider than joining atoms together. It is often used in organic chemistry to mean the joining of small molecules together to make bigger molecules, or converting one type of molecule into another. The term still applies to the joining of materials to form a single new compound. The alcohol methanol, for example, can be made by the joining together of hydrogen and carbon monoxide under suitable conditions of heat and pressure with a catalyst being present. The equation for this reaction is:

$$CO(g) + 2H_2(g) \rightarrow CH_3OH(l)$$

The opposite of synthesis is **decomposition**.

Answers to diagnostic test

If you score less than 80%, then work through the text and re-test yourself at the end by using this same test. If you still get a low score, then re-work the unit at a later date.

1 Binary compounds consist of <u>two</u> <u>elements</u> which are chemically <u>combined</u>. Their names end with 'ide'.   **(4)**

2 a) Magnesium oxide;
b) zinc sulfide;
c) silicon chloride;
d) hydrogen sulfide;
e) lead bromide.
(2 marks each)   **(10)**

3 Coal contains carbon (1) hydrogen (1) and sulfur (1). Combustion is combination with oxygen (1). Carbon is oxidised (1) to carbon dioxide (1). Hydrogen is oxidised to water (1). Sulfur is oxidised to sulfur dioxide (1).   **(8)**

22 Marks   (80% = 18)

FURTHER QUESTIONS (NO ANSWERS GIVEN)

One story about the discovery of copper goes like this:

In ancient times a party of hunters stopped for the night. They built a fire hearth with some of the green-coloured rocks containing copper carbonate which they found around them. They piled wood (which, when heated forms charcoal, which is almost pure carbon) inside this and lit a huge fire for warmth and light. They took it in turns to keep the fire burning through the night. Sometimes the flames were tinged with green colour. In the morning one of them noticed that a reddish-brown metal had oozed from the rocks and solidified.

Why? Suggest the possible reactions that had taken place.

Unit 3.2
Decomposition – The Breaking Up of Compounds

Starter check

What you need to know before you start the contents of this unit.

The contents of Unit 3.1. Also the nature and properties of ionic compounds.

Aims **By the end of this unit you should understand that:**

- Compounds can be broken up.
- The breaking up of a compound is called decomposition.
- Decomposition can often be brought about by heating.
- Decomposition can often be brought about by electricity.

Diagnostic test

Try this test at the start of the unit. If you score more than 80%, then use this unit as a revision for yourself and scan through the text. If you score less than 80%, then work through the text and re-test yourself at the end by using this same test.

The answers are at the end of the unit.

1 Which of the following, from the evidence presented, **must** be compounds?

a) Substance A: on heating goes from white powder to yellow and on cooling returns to a white powder.

b) Substance B: a green powder, which on heating turns black and gives off a colourless gas.

c) Substance C: a white powder, which on melting and having electricity passed through it, produces a colourless gas and a silvery metal.

d) Substance D: a white powder, which does not conduct electricity either in the solid or the molten state and is unaffected by the electric current. **(4)**

2 What sort of compound is each of the following?

   a) Substance A: decomposes on heating to give carbon
     dioxide and a metallic oxide. **(2)**
   b) Substance B: decomposes on heating to give a nitrite
     of the element plus oxygen gas. **(4)**
   c) Substance C: decomposes on heating to give nitrogen
     dioxide, oxygen and a metallic oxide. **(4)**
   d) Substance D: decomposes on heating to give sulfur
     trioxide and a metallic oxide. **(4)**

3 What is the meaning of the following terms?

   a) Electrolysis **(3)**
   b) Electrolyte **(2)**
   c) Cathode **(2)**
   d) Anode **(2)**

4 At which electrode would you expect a metal to be deposited
during electrolysis of a molten metallic chloride? **(2)**

29 Marks   (80% = 23)

3.2.1 Thermal Decomposition

In 1774, Joseph Priestley took delivery of a powerful magnifying glass (see Figure 3.2.1).

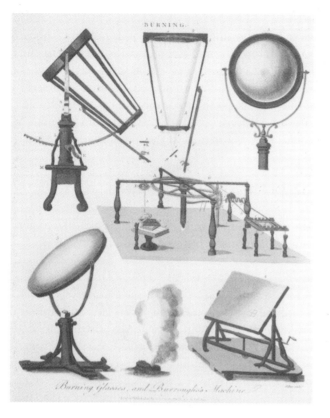

Figure 3.2.1
Priestley's 'burning glass'

He used it to focus the Sun's rays on red calx (the ancient name for oxide) of mercury. The heat caused a silvery liquid and a colourless gas to be produced. The silvery liquid was mercury. The colourless gas was something that had never been seen in a pure state before. It was oxygen, although that's not the name that Priestley gave it – he called it 'dephlogisticated air'. It was the French chemist Antoine Lavoisier who gave it its modern name. Red calx of mercury is nowadays called mercury(II) oxide. It is made of mercury and oxygen chemically combined together. It is not a particularly stable compound – the two elements split apart quite easily when heated and the compound decomposes. This type of reaction is called thermal decomposition because it's brought about by heating. The weaker the bonds holding the elements together in a compound, the more easily decomposed it will be. Mercury is not a particularly reactive metal; it isn't all that keen to combine with oxygen in the first place, and it splits apart from it quite easily. Sodium, on the other hand, is a very reactive metal. It is very keen to combine, and once it gets hold of some oxygen it won't let go of it easily. In other words, sodium oxide is a very stable compound, an ionic compound and it isn't easily decomposed. The more reactive elements are, the more stable their compounds are likely to be. (Remember the reactivity series – see Section 2.4.4.)

When compounds decompose, they don't necessarily split up into their constituent elements but sometimes into smaller compounds. In the Yorkshire and Derbyshire Dales a lot of limestone is quarried. Much of this is used as a construction and building material, but a lot is thermally decomposed in lime kilns.

Limestone is, chemically, calcium carbonate; it is made up of calcium, carbon and oxygen. When it is heated, it doesn't split up into these three elements, but into two simpler compounds – calcium oxide and carbon dioxide.

$$CaCO_3(g) \rightarrow CaO(s) + CO_2(g)$$

'Quicklime' is so called because it is pretty lively – and dangerous too. (Think of the phrase 'the quick and the dead'.) When water is added it fizzes, spits, crumbles, swells and gets very hot. It becomes 'slaked lime' (because 'its thirst is *slaked*'!). It is now what farmers call 'lime' when they put it on fields to neutralise acid soil. 'Limestone' is called limestone because it supplied lime!

Calcium oxide has the ancient name of quicklime. Most carbonates, when heated, undergo thermal decomposition to form the metal oxide and carbon dioxide. Sodium and potassium carbonate, on the other hand, don't decompose at all when heated, because sodium and potassium are very reactive metals, so their compounds are very stable.

Nitrates are made up of a metal, nitrogen and oxygen. Most nitrates undergo thermal decomposition to the metal oxide, oxygen and nitrogen dioxide:

$$2Cu(NO_3)_2(s) \rightarrow 2CuO(s) + O_2(g) + 4NO_2(g)$$

Sodium and potassium nitrates, however, give off oxygen as the only gas on heating, and leave the corresponding nitrite.

$$2KNO_3(s) \rightarrow 2KNO_2(s) + O_2(g)$$

This reaction can also be looked at in terms of oxidation and reduction. Look particularly at the N in KNO_3, its oxidation state is +5; as compared with the N in KNO_2 where it is +3. Two nitrogen atoms have dropped from +5 to +3 – they have been reduced; at the same time two oxygen atoms have gone from −2 to 0 – they have been oxidised.

Mercury(II) nitrate decomposes first into mercury(II) oxide which then decomposes further into mercury metal and the two gases oxygen and nitrogen dioxide:

$$2Hg(NO_3)_2(s) \rightarrow 2Hg(s) + 2O_2(g) + 4NO_2(g)$$

Similarly, some sulfates and hydroxides will decompose on heating. In fact, sulfuric acid was discovered by Arabian chemists in the Middle Ages by heating copper(II) sulfate crystals [or 'hydrated copper(II) sulfate', which they called 'blue vitriol']. They heated it in a clay pot and collected the condensed vapours. Firstly, it gave off water vapour, which cooled to condense as liquid water.

$$CuSO_4 \cdot 5H_2O(s) \rightarrow CuSO_4(s) + 5H_2O(l)$$

The anhydrous copper(II) sulfate then decomposed on further strong heating to copper(II) oxide and sulfur trioxide.

$$CuSO_4(s) \rightarrow CuO(s) + SO_3(g)$$

The sulfur trioxide dissolved in the water to form sulfuric acid.

$$SO_3(g) + H_2O(l) \rightarrow H_2SO_4(aq)$$

3.2.2 Electrolysis of Molten Compounds

Towards the end of the eighteenth century, it became possible to produce a continuous electric current. In 1808, Humphry Davy discovered that if you passed electricity through melted potash (potassium hydroxide) and put a positive and a negative electrode into the molten compound, the potash was decomposed into oxygen and a silvery grey metal he called 'potassium'. Up until that point, potash had been thought of as an element, as there was no apparent way of splitting it up into something simpler. Now it was known that it was composed of potassium and oxygen.

Similar experiments were performed on other substances which until then had been thought of as elements. Soda (sodium hydroxide) was melted and had electricity passed through it; it split up into sodium and oxygen.

The general rule was found to be that if a compound contained a metal and it could be melted, electricity could be passed through it and it would split up; the metal it contained was always attracted to the negative electrode, or cathode. The non-metal it contained was always attracted to the positive electrode, or anode. The process of passing an electric current through a molten compound and thus decomposing it was known as **electrolysis** (any word ending in 'lysis' indicates that splitting up of some kind is involved). The positive electrode in electrolysis is known as the anode and the negative electrode as the cathode (see Figure 3.2.2).

> An **electrolyte** is a substance which, when molten or in solution, conducts electricity and is broken up while doing so.

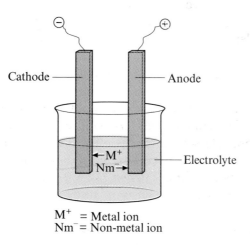

M^+ = Metal ion
Nm^- = Non-metal ion

Figure 3.2.2
Electrolysis

During electrolysis, the positive (usually metal) ions travel to the cathode; the negative ions travel to the anode. When they reach the electrodes, the ions are discharged, *i.e.* they lose their charge. The negative ions give electrons away and become neutral; the positive ions take in electrons and become neutral.

e.g. If molten lead bromide is electrolysed, the positive lead ions are discharged at the cathode thus:

$$Pb^{2+}(aq) + 2e^- \rightarrow Pb \text{ metal}(s)$$

In this process the lead ions have gained electrons; they have been reduced (remember OILRIG – see Section 2.5.1).

Similarly, the negative bromide ions are discharged at the anode; the bromide ions have lost electrons; they have been oxidised (remember OILRIG – see 2.5.1).

$$2Br^-(aq) \rightarrow Br_2(aq) + 2e^-$$

Note that in any electrolysis, reduction occurs at the cathode (CaRe – Cathode Reduction), so oxidation must occur at the anode.

3.2.3 Electrolysis of Aqueous Solutions

Solutions of compounds can also be electrolysed, but here the situation is more complicated. This is because there are extra ions present from the water in which the compound is dissolved.

Water contains hydrogen and hydroxide ions, because a few of the water molecules ionise.

$$H_2O(l) \rightleftharpoons H^+(aq) + OH^-(aq)$$

If you electrolyse a solution of salt, NaCl, for example, there will be two positive ions present, hydrogen ions (H^+) from the water and sodium ions (Na^+) from the salt. Both of these travel to the cathode. Only the hydrogen ions are discharged, however, because it takes too much energy to force a sodium ion to accept an electron. Electron gain is also called reduction, the oxidation number has gone down from +1 to 0.

$$2H^+(aq) + 2e^- \rightarrow H_2(g)$$

Although pure water only contains a very small number of hydrogen ions per litre, as soon as they are discharged more water ionises, so the electrolysis can continue. Note, incidentally, that hydrogen is the only non-metal to be produced at the cathode during electrolysis – usually non-metals are produced at the anode because they form negative ions. Bubbles of hydrogen are seen coming from the cathode.

In the same way there will be two negative ions present, hydroxide from the water and chloride from the salt; both travel to the anode. What happens when they get there depends on the concentration of the solution; if it is very dilute, it will be almost only the hydroxide ions that are discharged:

$$4OH^-(aq) \rightarrow 2H_2O(l) + O_2(g) + 4e^-$$

and oxygen will be given off. The more concentrated the solution, the more chlorine will be discharged:

$$2Cl^-(aq) \rightarrow Cl_2(g) + 2e^-.$$

Usually, a mixture of oxygen and chlorine is given off.

In all of these reactions the compounds must be ionic in nature and so contain ions. When electricity is passed through the melted compound or the solution, the ions move because they are attracted to the electrode which has opposite charge to their own. Remember this migration can only happen when the ions are free to move. They cannot migrate in the solid state.

The electricity causes the compound's ions to separate and so they are decomposed. The compound splits up into its components.

Answers to diagnostic test

If you score less than 80%, then work through the text and re-test yourself at the end by using this same test. If you still get a low score, then re-work the unit at a later date.

1 Substances B and C (2 points for each, knock off 1 point if you've put A or D) **(4)**

2 a) A carbonate; **(2)**

   b) a nitrate of potassium or sodium; **(4)**

   c) a nitrate other than of sodium or potassium; **(4)**

   d) a sulfate other than of potassium or sodium. **(4)**

3 a) The passage of an electric current (1) through a compound (1) with resulting decomposition (1). **(3)**

   b) A compound which, when in solution or melted, conducts electricity (1) with resulting decomposition of the compound (1). **(2)**

   c) The negative (1) electrode (1) during electrolysis. **(2)**

   d) The positive (1) electrode (1) during electrolysis. **(2)**

4 The cathode. **(2)**

29 Marks   (80% = 23)

FURTHER QUESTIONS (NO ANSWERS GIVEN)

1 Aluminium is the commonest metal in the Earth's crust, and yet up to the latter part of the nineteenth century it was very expensive, and knives, forks and spoons of aluminium were only available to the very rich. This was because it is so reactive that it couldn't be extracted from its oxide ore by the traditional method of reducing with carbon.

Charles Hall, in the USA, found a way of extracting aluminium cheaply enough to enable it to be widely used. Why are aluminium extraction plants often found in regions where there are mountains and lots of running water and cheap hydro-electricity? Write out the possible process taking place.

2 Aluminium oxide has a very high melting point, so a more easily melted compound of aluminium, sodium hexafluoroaluminate(III), called cryolite (Na_3AlF_6), was added. What will be the effect on the melting point of aluminium oxide of adding this impurity?

3 Aluminium is now a very common metal, but it is quite reactive and yet it is used for window frames, milk bottle covers, metal foil. How does this reactive metal become so relatively 'unreactive' when exposed to the air?

Unit 3.3
Exchange Chemical Reactions

Starter check

What you need to know before you start the contents of this unit.

You should understand what is meant by 'chemical reactions' and 'ions', and have a basic knowledge of what is meant by the words 'acid', 'electrolysis' and 'oxidation'.

Aims **By the end of this unit you should understand that:**

- Elements can 'swap partners' in some chemical reactions.
- A more reactive element can displace a less reactive one from its compounds.
- Neutralisation of an acid is an example of an exchange reaction.
- The reaction of a carbonate with an acid is an example of an exchange followed by a decomposition.

Diagnostic test

Try this test at the start of the unit. If you score more than 80%, then use this unit as a revision for yourself and scan through the text. If you score less than 80%, then work through the text and re-test yourself at the end by using this same test.

The answers are at the end of the unit.

1 Complete the following word equations by either writing the names of the products, or 'no reaction' if you think there will be no reaction:

a) Magnesium + copper sulfate →	(2)
b) Lead + zinc nitrate →	(2)
c) Sodium hydroxide + sodium nitrate →	(2)
d) Potassium hydroxide + zinc sulfate →	(2)
e) Aluminium + iron oxide →	(2)
f) Zinc + sulfuric acid →	(2)
g) Copper + hydrochloric acid →	(2)
h) Sodium hydroxide + nitric acid →	(2)
i) Calcium carbonate + hydrochloric acid →	(3)
j) Magnesium + water →	(2)

21 Marks   (80% = 17)

3.3.1 Displacement

Displacement reactions are characterised by one element displacing another out of its compound.

Iron used to be extracted from its ore, haematite, which is chemically iron(III) oxide, by heating the ore with charcoal. Nobody really knows how the first metals were extracted from their oxides, thousands of years ago, but it could have been by an accidental reduction of the oxide rock being heated in a charcoal fire and then the liquid metal solidifying. The wood and charcoal which were burning contain carbon, and carbon is more reactive towards oxygen than iron. Because it had a greater affinity for the oxygen, the carbon (given the energy from the fire) can take up the oxygen and leave iron on its own. Another way of saying this is that carbon can displace iron from its oxide. (Remember the reactivity series of Unit 2.4.) The carbon combines with the oxygen out of the iron oxide and keeps it for itself, the carbon is oxidised (*i.e.* picks up oxygen). The iron oxide loses oxygen to the carbon. The iron oxide is reduced. (Reduction is the loss of oxygen.) Whenever one element is reduced (loses oxygen), another must be oxidised (gain oxygen). This, as with other types of chemical reactions, can be classified in different ways depending on what you are looking at; here the reaction can be considered as a displacement reaction or a redox reaction.

This reaction can be scrutinised for oxidation number changes. Iron goes from +3 to 0 and so has been reduced, and the carbon is oxidised from zero oxidation number to +2 in carbon monoxide.

$$Fe_2O_3(s) + 3C(s) \rightarrow 2Fe(s) + 3CO(g)$$

(Actually, in the blast furnace it is carbon monoxide which does the greatest amount of the reduction.)

You can arrange the metals in order of reactivity. The elements at the top are the most reactive. Anything nearer the top will displace anything lower down from its compounds. (Note that you should never attempt a displacement reaction with any of the top five elements – it is far too dangerous!)

This list is called a reactivity series, which was discussed in detail in Unit 2.4.

To show that the elements nearer the top will displace the elements lower down from its compounds, when you dip some copper wire into a solution of silver nitrate, crystals of silver will start to grow on the copper wire because the copper, being more reactive than silver, will displace silver from the silver nitrate. The solution will go blue, the usual colour for copper compounds in solution (Figure 3.3.1).

$$Cu(s) + 2AgNO_3(aq) \rightarrow Cu(NO_3)_2(aq) + 2Ag(s)$$

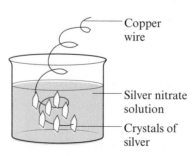

Figure 3.3.1
Copper displaces silver from
silver nitrate solution

The reverse will not work. Silver will not displace copper from copper sulfate solution, and nothing lower down the series will displace an element higher up the series. So copper will never displace hydrogen gas from a solution of a dilute acid, but zinc will.

The majority of the elements in the list are metals, but carbon and hydrogen are included because of the importance of carbon in extracting metals, and because of the importance of metal/acid reactions.

Any metal which is below carbon (coke) in the series can be displaced by carbon from its oxide.

Any metal above carbon (such as aluminium) cannot be extracted with coke.

Sodium can only be extracted by electrolysing molten salt (sodium chloride). See Section 3.2.2 for the explanation of electrolysis. The top four metals in the reactivity series shown here were not discovered until electrolysis of their molten compounds became possible early in the nineteenth century (see Section 3.2.1).

Non-metals also vary in reactivity. In, for example, Group VII of the periodic table (the halogens), chlorine is more reactive than iodine. If you pass chlorine gas into potassium iodide solution, the solution goes brown because the chlorine displaces iodine from the potassium iodide. Iodine in KI has an oxidation number of -1 and this rises to 0 in the elemental iodine, so iodine has been oxidised. The opposite occurs with chlorine.

$$2KI(aq) + Cl_2(aq) \rightarrow 2KCl(aq) + I_2(aq)$$

3.3.2 Acids and Metals

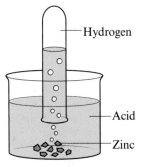

Figure 3.3.2
Zinc displaces hydrogen from the acid

All acids contain hydrogen. You can see from the reactivity series that zinc is more reactive than hydrogen, because zinc is above hydrogen in the series. If you add zinc to hydrochloric acid solution, it will displace hydrogen from the acid and end up joined with the chloride ion to produce a salt (Figure 3.3.2).

$$Zn(s) + 2HCl(aq) \rightarrow ZnCl_2(aq) + H_2(g)$$

Metals below hydrogen in the reactivity series cannot displace hydrogen from an acid. So copper, silver and gold are unreactive with most dilute acids and do not give off hydrogen gas.

When the hydrogen from an acid is displaced by a metal, the acid is neutralised because it is the hydrogen ions that make the acid an acid. These are given off as hydrogen gas and so have been lost to the solution.

In terms of electrons, metals lose electrons to hydrogen ions, which are discharged and become hydrogen gas. So, for example:

$$Zn(s) \rightarrow Zn^{2+}(aq) + 2e^-$$
$$2H^+(aq) + 2e^- \rightarrow H_2(g)$$

or, overall:

$$Zn(s) + 2H^+(aq) \rightarrow Zn^{2+}(aq) + H_2(g)$$

In general:

$$Metal + Acid \rightarrow Salt + Hydrogen$$

3.3.3 Ion Exchange Reactions

The more reactive a metal is, the more likely its compounds are to be soluble in water. Most carbonates, for example, are insoluble in water, but sodium and potassium carbonate are both soluble. All the other metal carbonates are insoluble in water.

All metal nitrates are soluble in water.

If you add a solution of sodium carbonate to a solution of lead nitrate, you will see a solid white precipitate of lead carbonate being produced – the mixture will look cloudy because the particles of solid lead carbonate block light from passing through the liquid. What has happened is that the lead and sodium have 'swapped partners' – an exchange reaction has occurred.

e.g. Lead nitrate + sodium carbonate → lead carbonate
    + sodium nitrate

$$Pb(NO_3)_2(aq) + Na_2CO_3(aq) \rightarrow PbCO_3(s) + 2NaNO_3(aq)$$

Almost any insoluble salt can be made by mixing a solution of the metal nitrate (which is bound to be soluble) with a solution of the sodium salt of the acid (ditto).

Similarly, if you add sodium hydroxide solution to copper sulfate solution, you get a precipitate of copper hydroxide, which is insoluble. (The only common soluble hydroxides are sodium, potassium and, to a slight extent, calcium.)

$$CuSO_4(aq) + 2NaOH(aq) \rightarrow Cu(OH)_2(s) + Na_2SO_4(aq)$$

Some general characteristics of salts (*i.e.* compounds containing metal or ammonium ions with ions from acids) are:

Soluble substances
All sodium, potassium and ammonium salts are soluble in water.
All nitrates are soluble.
Most chlorides are soluble, except silver and lead chlorides.
Most sulfates are soluble, except lead and barium.
All acids are soluble.

Insoluble substances
All carbonates are insoluble except for those of sodium, potassium and ammonium.
All common lead salts are insoluble, except lead nitrate.
All common hydroxides and oxides are insoluble, except for those of sodium and potassium.

3.3.4 Neutralisation of Acids

Metal oxides react with acids in exchange reactions. All acids contain hydrogen. Obviously all metal oxides contain oxygen. When a metal oxide is added to an acid, the hydrogen from the acid swaps partners with the metal from the oxide and joins up with the oxygen to form water.

e.g. Copper oxide + sulfuric acid → copper sulfate + water

$$CuO(s) + H_2SO_4(aq) \rightarrow CuSO_4(aq) + H_2O(l)$$

The acid loses its hydrogen, so it is neutralised. Anything that neutralises an acid is called a base. All metal oxides are basic oxides, because they neutralise acids.

A soluble base is called an alkali. Most metal oxides don't dissolve. The ones that do dissolve react with the water to form a hydroxide. Only sodium, potassium and (slightly) calcium hydroxides are soluble.

e.g. Sodium oxide + water → sodium hydroxide

$$Na_2O(s) + H_2O(l) \rightarrow 2NaOH(aq)$$

Sodium hydroxide is an alkali. When added to an acid, the hydroxide ion from the sodium hydroxide combines with the hydrogen ion from the acid to form water, and the remaining acid ion combines with the sodium to form a salt. To put it in its simplest terms, the hydrogen ions from an acid neutralise the hydroxide ions from an alkali by combining with them to form water molecules. The salt is simply the ions which are left in solution, and to get the solid salt you simply have to evaporate off the water.

e.g. Sodium hydroxide + sulfuric acid → sodium sulfate + water

$$2NaOH(aq) + H_2SO_4(aq) \rightarrow Na_2SO_4(aq) + 2H_2O(l)$$

Metal carbonates react with acids in exchange reactions:

e.g. Sodium carbonate + sulfuric acid → sodium sulfate + hydrogen carbonate

$$Na_2CO_3(s) + H_2SO_4(aq) \rightarrow Na_2SO_4(aq) + H_2CO_3(aq)$$

Hydrogen carbonate is carbonic acid, which spontaneously decomposes to water and carbon dioxide:

Carbonic acid → water + carbon dioxide

$$H_2CO_3(aq) \rightarrow H_2O(l) + CO_2(g)$$

The acid is again neutralised because it loses its hydrogen ions.
The general equations are:

Basic oxide + Acid → Salt + Water
$$O^{2-}(aq) + 2H^+(aq) \rightarrow H_2O(l)$$

Alkali + Acid → Salt + Water
$$OH^-(aq) + H^+(aq) \rightarrow H_2O(l)$$

Carbonate + Acid → Salt + Water + Carbon dioxide
$$CO_3^{2-}(aq) + 2H^+ \rightarrow H_2O(l) + CO_2(g)$$

3.3.5 Summary of Chemical Reactions of Acids and Alkalis

Acids
All acids contain hydrogen which supplies hydrogen ions when in solution.

Acids generally liberate hydrogen gas when reacting with most metals (see the reactivity series).

Acids liberate carbon dioxide from any metallic carbonate.

Acids can be neutralised by bases and alkalis to form salts and water.

Acids have a pH in solution below 7 and turn litmus indicator or Universal Indicator red.

Acids are usually compounds of non-metals, *e.g.* hydrochloric acid HCl, sulfuric acid H_2SO_4, nitric acid HNO_3.

Alkalis
Alkalis are soluble bases. Bases are metallic oxides or hydroxides.

Alkalis give OH^- ions in aqueous solution.

Alkalis (and bases) react with acids to form salts and water.

Alkalis have pH values of above 7 and turn litmus or Universal Indicator solution blue.

Answers to diagnostic test

If you score less than 80%, then work through the text and re-test yourself at the end using this same test. If you still get a low score, then re-work the unit at a later date.

1 a) magnesium sulfate (1) + copper (1) **(2)**
 b) no reaction **(2)**
 c) no reaction **(2)**
 d) potassium sulfate (1) + zinc hydroxide (1) **(2)**
 e) aluminium oxide (1) + iron (1) **(2)**
 f) zinc sulfate (1) + hydrogen (1) **(2)**
 g) no reaction **(2)**
 h) sodium nitrate (1) + water (1) **(2)**
 i) calcium chloride (1) + water (1) + carbon dioxide (1) **(3)**
 j) magnesium oxide/hydroxide (1) + hydrogen (1) **(2)**

21 Marks (80% = 17)

FURTHER QUESTIONS (NO ANSWERS GIVEN)

1 Milk of magnesia (magnesium oxide) is often used to treat acid indigestion, or heartburn. Why?
2 Slaked lime is often added to acid soil to increase fertility. Why?
3 How could the following be made?

a) Lead nitrate from lead metal.
b) Lead carbonate from lead metal.
c) Magnesium oxide from magnesium sulfate solution.

Unit 3.4
Some Further General Characteristics of Chemical Reactions

Starter check

What you need to know before you start the contents of this unit.

You will need to understand about chemical reactions, and a review of Units 3.1, 3.2 and 3.3 would be helpful.

Aims **By the end of this unit you should understand that:**

- Reactions involve energy changes.
- Reactions involve other changes including oxidation and reduction.
- Some of these changes can be measured.

Diagnostic test

Try this test at the start of the unit. If you score more than 80%, then use this unit as a revision for yourself and scan through the text. If you score less than 80%, then work through the text and re-test yourself at the end by using this same test.

The answers are at the end of the unit.

1 What do the terms 'exothermic' and 'endothermic' mean?   **(4)**
2 What does redox mean?   **(2)**
3 What is 'cracking' in terms of crude oil? Is it exo- or endothermic? (You should be as detailed as possible, and use correct scientific/technical terms.)   **(10)**
4 The earliest electrical cells that were made contained copper and zinc plates separated by brine-soaked cardboard discs. How did they produce electricity?   **(9)**
5 Give one example from nature of a chemical reaction that produces light directly.   **(2)**

27 Marks   (80% = 22)

3.4.1 Heat Energy

Everything is made up of atoms, molecules or ions. At any temperature above Absolute Zero (-273 °C), these particles are moving. Before one substance can react with another, their particles must actually collide with each other.

Take, for example, the reaction between hydrogen and oxygen. When they react, water is produced.

Hydrogen + oxygen → water

$2H_2(g) + O_2(g) \rightarrow 2H_2O(l)$

The hydrogen burns in oxygen, and given the right mix of the two gases, it will explode quite violently. In the process, energy is released. The energy is in the form of heat, light and sound (if it explodes). The energy is given out to the environment, *i.e.* it is lost from the reaction mixture. Reactions that give out heat are called **exothermic** reactions (it's easy to remember this – you just need to think of 'exit' and 'out'). A complete study of the energy changes associated with chemical reactions is dealt with in Module 5.

Burning, or combustion, always gives out heat – a fact that humans have used for thousands and thousands of years to obtain heat energy. The burning of a fuel almost always involves combination with oxygen, or oxidation, and it is always exothermic.

Burning isn't the only exothermic reaction. There are, for example, emergency heating packs used by mountain rescue people that depend on an exothermic reaction. They contain two chemicals that react exothermically when the barrier between two containers of chemicals is broken. Isn't it dangerous carrying chemicals like that about? Not really, because the chemicals are iron and water and the reaction is rusting – but in this case a third chemical is added as a catalyst to speed up the reaction so that the heat is given out much more quickly than normal.

Reactions that take in heat energy are called endothermic reactions – these are not common.

All chemical reactions are either exo- or endothermic.

Some chemical reactions need a little bit of heat to get them going, then they continue on their own and give out heat, like initially applying a match to a candle: once it is lit it continues on its own.

3.4.2 Electricity

About two hundred years ago, Luigi Galvani noticed some frogs' legs twitching when he hung them on hooks from metal rails around his laboratory. The hooks and rail were made of different metals. Another scientist, called Alessandro Volta, worked out that this happened because the hooks were brass and the rail was steel. He decided that electricity could be produced by putting two different metals in a conducting solution, in this case the frogs' blood, and joining them by a wire. The legs were twitching because a very small amount of electricity was being produced by a chemical reaction between the metals, which was picked up by the frogs' legs when hung on them, and this made the muscles in the legs contract.

Suppose you dip a piece of zinc into copper sulfate solution. Zinc is more reactive than copper, so it can displace copper from its compounds (remember the reactivity series of Units 3.3 and 2.4):

Zinc + copper sulfate → zinc sulfate + copper

$$Zn(s) + CuSO_4(aq) \rightarrow ZnSO_4(aq) + Cu(s)$$

The zinc will become coated with metallic copper.

What is happening in terms of the particles concerned? Copper sulfate is an ionic compound, made up of copper (Cu^{2+}) and sulfate (SO_4^{2-}) ions. Zinc sulfate is an ionic compound, made up of zinc (Zn^{2+}) and sulfate (SO_4^{2-}) ions. The equation for the reaction can therefore be written like this:

Zinc + copper sulfate → zinc sulfate + copper

$$Zn(s) + Cu^{2+}(aq) + SO_4^{2-}(aq) \rightarrow Zn^{2+}(aq) + SO_4^{2-}(aq) + Cu(s)$$

In this equation, the same number of sulfate ions appear on each side, *i.e.* they start the reaction as sulfate ions and finish it still as sulfate ions. In other words, nothing happens to them – they are simply *spectator ions*, so they don't need to appear in the equation at all:

$$Zn(s) + Cu^{2+}(aq) \rightarrow Zn^{2+}(aq) + Cu(s)$$

So, zinc atoms are being converted to zinc ions, and copper ions are being converted into copper atoms. What does this involve?

Zinc atoms are electrically neutral, zinc ions are positively charged. For zinc atoms to be converted to zinc ions, they must lose electrons:

$$Zn(s) \rightarrow Zn^{2+}(aq) + 2e^-$$

Copper atoms are electrically neutral, copper ions are positively charged. For copper ions to be converted to copper atoms, they must gain electrons:

$$Cu^{2+}(aq) + 2e^- \rightarrow Cu(s)$$

In other words, in this reaction electrons are moving from the zinc to the copper.

Now suppose you dip a piece of zinc and a piece of copper into a solution of copper sulfate that will conduct electricity, so making a complete circuit. If you connect the two pieces of metal with a piece of wire, electrons will flow from the zinc to the copper *via* the wire – in other words, an electric current will be produced in the wire.

This means that *some chemical reactions can produce electrical energy* (which is the exact opposite of electrolysis, in which electrical energy causes chemical reactions).

The very first batteries were called Voltaic Piles, after Volta, who invented them. They consisted of alternate discs of copper and zinc, separated by cardboard soaked in salt solution (Figure 3.4.1).

Modern batteries operate on exactly the same principle, but using different materials. The most common sort have a zinc casing, which forms the negative pole; a central carbon rod, which forms the positive pole; and are filled with a paste of ammonium chloride, to let the ions flow.

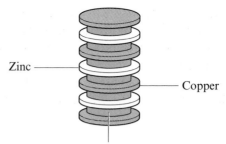

Zinc

Copper

Figure 3.4.1
A Voltaic Pile

Brine-soaked cardboard discs

People discovered how to produce current electricity about two hundred years ago, but parts of the animal kingdom, such as the electric eel, discovered it millions of years ago – these eels produce electrical energy *via* the reactions going on in their bodies.

Similarly, glow-worms and fireflies have chemical reactions in their bodies which can produce light energy. It is only recently that Man has found out how these reactions work and been able to reproduce them in the kind of 'glow tubes' they sell at fairs, *etc*. The important thing to remember is that these reactions produce light directly. Most of the ways we have of making light depend on producing a reasonable amount of heat (in a flame, for example) which gets something hot enough to start glowing. That would be no good in a glow-worm's body: they would burn up. The light they produce is made directly by the reaction, and only a very little heat is produced.

3.4.3 Redox Reactions

You have already come across some chemical reactions which involve the addition or subtraction of oxygen. Such reactions can also be classified as oxidation or reduction reactions.

In the old definition, oxidation is a process in which oxygen is added to an element or compound or, alternatively, one in which hydrogen is removed. The modern definition is that oxidation is a process in which there is electron loss from an element or compound (see Module 2), and it can also be described as the increase in the oxidation state of an element.

Oxidising agents (the chemicals that do the oxidation) are themselves reduced as they give up the oxygen or pick up the electrons.

Reduction is the opposite of oxidation, *i.e.* it is a loss of oxygen, gain of hydrogen or, in modern terms, a gain of electrons; and the reducing agent is itself oxidised as it picks up the oxygen, loses its hydrogen or loses its electrons. Reduction involves a decrease in the oxidation state of an element.

Whenever there is a process of oxidation there must be an accompanying process of reduction. It is always wise to say what has been oxidised/reduced to what. This topic is also covered in more detail in a different way when considering oxidation states in Unit 2.5.

Chemical reactions which involve Reduction and Oxidation are called REDOX reactions.

Example 1:
Copper oxide can react with hydrogen gas to make copper and water.

$$CuO(s) + H_2(g) \rightarrow Cu(s) + H_2O(l)$$

The copper oxide has been reduced (loss of oxygen) to copper. Hydrogen has gained oxygen and so has been oxidised to water. To put it in more modern terms, the oxidation state of the copper has fallen from $+2$ to 0 and that of hydrogen has increased from 0 to $+1$.

The oxidising agent is copper oxide which has itself been reduced, and the hydrogen is the reducing agent and has itself been oxidised.

Example 2:
Hydrogen sulfide gas when mixed with sulfur dioxide gas makes water and a deposit of sulfur. The balanced equation is shown.

$$2H_2S(g) + SO_2(g) \rightarrow 2H_2O(l) + 3S(s)$$

The oxidation state of the sulfur in the H_2S is -2; the oxidation state of the sulfur in the SO_2 is $+4$; the oxidation state of all the sulfur on the right-hand side of the equation is 0. So the sulfur from the H_2S has been oxidised, and the sulfur from the SO_2 has been reduced.

This unit often brings together some ideas and reactions that have appeared in other units in a different context. It should help to consolidate these ideas.

Answers to diagnostic test

If you score less than 80%, then work through the text and re-test yourself at the end using this same test. If you still get a low score, then re-work the unit at a later date.

1 An exothermic reaction <u>gives out</u> heat <u>to its surroundings</u>.
   An endothermic reaction <u>takes in</u> heat <u>from its surroundings</u>.   **(4)**

2 Redox is the term used when oxidation and reduction are
   simultaneously taking place in a chemical reaction.   **(2)**

3 Cracking is the conversion of <u>high boiling point fractions</u>
   from <u>crude oil</u> into <u>more valuable lower boiling point fractions</u>.
   The process is carried out by passing the <u>hot vapour</u> over a
   <u>catalyst</u>. The <u>larger molecules</u> <u>break down</u> to <u>form smaller</u>
   <u>ones</u>. It is a <u>thermal decomposition reaction</u>, which is
   <u>endothermic</u>.   **(10)**

4 <u>Zinc is more reactive than copper</u>. When placed in a <u>liquid</u>
   <u>that conducts</u> *i.e.* an electrolyte, <u>electrons</u> <u>flow or move</u>
   <u>through an externally connected wire</u> <u>from the zinc</u> <u>to the</u>
   <u>copper</u>. The <u>chemical reaction produces or gives out</u>
   <u>electrical energy</u>.   **(9)**

5 In the bodies of glow-worms and fireflies (for example).   **(2)**

27 Marks   (80% = 22)

FURTHER QUESTIONS (NO ANSWERS GIVEN)

1 You can buy 'glow sticks' as a way of producing light in an emergency. They are made of a clear, flexible plastic containing liquid. When you bend them another tube inside breaks and they start to produce 'heatless light'. How?

2 The general term for a reaction that takes in energy is endoergic. Chlorophyll is involved in a very important endoergic reaction. What is the reaction and what are the reagents and products?

3 Explain how the reactivity series can be considered as another way of explaining redox reactions.

4 Discuss the following reaction pathways in terms of redox reactions.
   Sulfur is burned in air to make sulfur dioxide. This gas is then mixed with more oxygen and passed over a hot catalyst to make sulfur trioxide, which is then dissolved in water to make sulfuric acid.

5 Ammonia can react with oxygen in the presence of a hot catalyst to make nitric oxide (NO) and water. The NO reacts with further oxygen to make nitrogen dioxide (NO_2) and this, when dissolved in water in the presence of oxygen, makes nitric acid. Write out the balanced equations, and explain the reactions in terms of redox processes.

Module 4

Quantitative Aspects of Chemical Reactions

Unit 4.1
Mathematical Concepts

Starter check

What you need to know before starting this unit.

Basic mathematical ideas.

Aims **By the end of this unit you should understand how to:**

- Convert an exponential and powers of ten into decimals.
- Use an exponential in calculations.
- Use your own calculator for calculations and 'logs'.
- Rearrange equations.
- Plot and work with graphs.

Diagnostic test

Try this test at the start of the unit. If you score more than 80%, then use this unit as a revision for yourself and scan through the text. If you score less than 80%, then work through the text and re-test yourself at the end using this same test.

The answers are at the end of the unit.

1 If the relationship between X and Y is given by equation below and you are given the value of X, rearrange the equation so that you can find a value for Y. **(2)**

$$X = \frac{3Y + 4}{6}$$

2 Change the number 0.000 003 97 to the format $n \times 10^{m}$. **(2)**

3 Find the logarithm to base 10 ($\log_{10}$) of the number in Question 2. **(2)**

4 Round up the number 3.868 131 025 808 9 to three significant figures. **(2)**

5 $2 \times 10^{6} \times 3 \times 10^{5} = ?$ **(2)**

10 Marks   (80% = 8)

4.1.1 Introduction

A small boy wearing a red scarf and football shirt is running along a road in Manchester. He meets an old man. The small boy asks, 'Can you tell me how to get into Manchester United?' The old man replies, 'You need practice, lad, lots of practice!' To understand and use some aspects of chemistry, you need to be able to use simple maths, and that means practice. Some people are put off by maths, but in this unit, and in chemistry in general, the maths can be based on just a few ideas. We will work through these ideas in this text. It does require you to follow the logic step by step, and practise with the given examples. In general, the maths required for this course is all covered in the GCSE maths course, with one possible exception – logarithms. But don't panic – although a short theory of logs is given as an extension topic at the end of this unit, with modern calculators it is not necessary to understand how logs work before you can use them.

GENERAL POINTS

Always read the question carefully to decide:

a) what information is provided;
b) what exactly is to be worked out.

Think about the relationship between (a) and (b). In general, any equation will only have one unknown. If information, *e.g.* a value for a particular constant, has been provided, then it will be needed. If you haven't used it, you have probably used the wrong relationship.

When you have an answer, think about what approximate answer you could expect to get. Does your answer look sensible? For example, the pH of an acid will be below 7, so if your answer is 113, think again!

4.1.2 Standard Form (Exponential) Numbers and Powers of 10

When an old friend of ours was writing a note asking for a favour, he usually finished the note with the phrase 'Thanks a 10^6'. That expression, 10^6, spoken as 'ten to the sixth' or ' ten to the power six', is the scientific way of saying 'one million', and is known as an **exponential number**, or a number written in **standard form**, *i.e.* one expressed in powers of 10.

An exponential number, or a number written in standard form, must always be written as:

$A \times 10^n$

(*A* must always be between 1 and 10)
(*n*, the number to which ten is raised, is the number of places the decimal point 'moves')

Ten raised to the power 6 is equal to 10 multiplied by itself 6 times, $10 \times 10 \times 10 \times 10 \times 10 \times 10$. Written as an ordinary number it is 1 followed by 6 noughts, 1 000 000.

One thousand can be written as 1000 or 1×10^3. So 1 is multiplied by 10 three times, $10 \times 10 \times 10$. To get this, we move the decimal place to the left three times:

$$1000 = 1\ 0\ 0\ 0 = 1 \times 10^3$$
$$3\ 2\ 1$$

Generally we can say that 10 raised to the power n, or 10^n, is 1 followed by n noughts. Using numbers in standard form is very convenient when dealing with very large numbers.

e.g. 1×10^{20} is 1 followed by 20 noughts, or 100 000 000 000 000 000 000, or 1 hundred million, million million. (And one of the most important numbers in science, the Avogadro constant, is 6000 times bigger than that! – see Section 4.3.1.)

What about converting the number 5 000 000 to standard form?

5 000 000 is $5 \times 1\,000\,000$, so in this case we still move the decimal place to the left until we have only one figure before the decimal point, and then count up the number of times we have moved it. In this case we have to move it six times, so the answer is 5×10^6:

$$5\,000\,000 = 5 \times 1\,000\,000 = 5 \times 10^6$$

If the power to which 10 has to be raised is a negative number, *e.g.* 10^{-1}, then the power shows how many times we are dividing by 10, *i.e.* 10^{-1} is 1/10 or 0.1.

So, 1×10^{-6} is 1 divided by 10 six times, moving the decimal point to the left six times:

$$0\ 0\ 0\ 0\ 0\ 1. \qquad 1 \times 10^{-6} \text{ then becomes } 0.000\,001$$
$$6\ 5\ 4\ 3\ 2\ 1$$

$$10^{-6} = 1/1\,000\,000 = 1 \text{ millionth} = 0.000\,001$$

RULES FOR WORKING WITH EXPONENTIAL NUMBERS OR NUMBERS IN STANDARD FORM

1   When the power to which 10 has to be raised is positive, the number will be greater than 1, *e.g.* 10^2 is 100.
2   When the power to which 10 has to be raised is negative, the number will be less than 1, *e.g.* 10^{-2} is 1/100 or 0.01.
3   To convert a written number to a number in standard form:
   a)   move the decimal point until it appears after the first non-zero number.

   e.g. $0\ .\ 0\ 1$ two places moved; 0.01 becomes 10^{-2}
   $$1\ 2$$

   b)   Count the number of places you have moved it. That is the power to which the 10 is raised. If you have moved the decimal point to the right, then the power is a negative number. Thus moving two places to the right, the number 0.01 becomes 10^{-2}.

NOTE: 10 raised to the power $+n$ (10^n) will give 1 followed by n noughts, but if 10 is raised to the power $-n$ (10^{-n}) there will only be $(n-1)$ noughts between the decimal point and the 1.

Example 1:
Convert 1 000 000 000 to standard form.

Answer: to give a number in the format '1 ×' we must move the decimal point 9 places to the left, so the answer is 1×10^9.

Example 2:
Convert 725 000 to standard form.

Answer: move the decimal point so the number reads 7.25. To get this, we have moved the decimal point 5 places to the left. The answer is 7.25×10^5.

Example 3:
Convert 0.000 005 67 to standard form.

Answer: move the decimal point to the right so the number reads 5.67. To get this we have moved the decimal point 6 places. The answer is 5.67×10^{-6}.

Example 4:
Convert 8.92×10^{12} to ordinary numbers.

Answer: the decimal point has to be moved 12 times to the right. The answer is 8 920 000 000 000.

Example 5:
Convert 6.24×10^{-3} to ordinary numbers.

Answer: the decimal point has to be moved three times to the left. The answer is 0.006 24.

4.1.3 Calculations with Numbers in Standard Form

ADDING AND SUBTRACTING

If the numbers to be added have the same power of 10, then the addition is just like adding 'normal' numbers:

Example 1: $6.24 \times 10^4 + 3.79 \times 10^4$ can be written as
$(6.24 + 3.79) \times 10^4 = 10.03 \times 10^4 = 1.003 \times 10^5$

If the numbers to be added have different powers of 10, then express all the numbers to be added or subtracted to the same power of 10. Then add or subtract decimals in the normal manner.

Example 2: What is $6.24 \times 10^4 + 3.79 \times 10^5$?
The number 3.79×10^5 can be written as 37.90×10^4 (move the decimal point 1 place to the right so the number can be '$\times 10^4$').

$$6.24 \times 10^4 + 37.90 \times 10^4 = (6.24 + 37.90) \times 10^4$$
$$= 44.14 \times 10^4$$
$$= 4.41 \times 10^5$$

Example 3: $3.65 \times 10^{-6} + 8.17 \times 10^{-5}$.

In the same way as above, 3.65×10^{-6} can be written as 0.365×10^{-5} (move the decimal point 1 place to the left so the number can be '$\times 10^{-5}$')

$$3.65 \times 10^{-6} + 8.17 \times 10^{-5} = 0.365 \times 10^{-5} + 8.17 \times 10^{-5}$$
$$= (0.365 + 8.17) \times 10^{-5}$$
$$= 8.535 \times 10^{-5}$$

MULTIPLICATION AND DIVISION

In multiplication, multiply the numbers and decimals in the usual way, but add the powers to which the 10 is raised.

e.g. add: $3 + 4 = 7$

$$10^3 \times 10^4 = 10^{(3+4)} = 10^7$$

e.g. add: $6 + 5 = 11$

$$2 \times 10^6 \times 3 \times 10^5 = (2 \times 3) \times (10^6 \times 10^5) = 6 \times 10^{(6+5)} = 6 \times 10^{11}$$

In division, divide the numbers as normal and subtract the power to which the 10 is raised.

e.g. $\dfrac{10^4}{10^2} = 10^{(4-2)} = 10^2$

e.g. $\dfrac{3 \times 10^6}{2 \times 10^5} = 3/2 \times 10^{(6-5)} = 1.5 \times 10^1 = 15$

e.g. $\dfrac{4.5 \times 10^3}{1.4 \times 10^5} = 4.5/1.4 \times 10^{(3-5)} = 3.2 \times 10^{-2}$

4.1.4 Using a Calculator

Get to know your own calculator; each calculator is slightly different in its mode of operation. Some of the most frequently used calculations covered by these exercises are:

WORK WITH NUMBERS WRITTEN IN STANDARD FORM

On our calculator the key sequence would be as follows. It is essential to try these out. Your calculator may be different. Read the instructions!

Example: 6.1×10^{-5}

	Key sequence
• Numerals: press keys for 6.1	[6], [.], [1]
• One press on the exponential key (this means 'times a power of 10')	[EXP]
• Numeral: 5 (this means '10^5')	[5]
• One press on the $+/-$ key gives the answer	[$+/-$]
• The calculator display will show 6.1×10^{-5}	

- Some calculators will display 6.1^{-05}
- When writing out the answer you **must** put in the
  10 because 6.1×10^{-5} is not the same as 6.01^{-5}

It is not necessary to input '$\times 10$' manually; if you do this, it will result in the answer being out by a factor 10.

WORK WITH LOGARITHMS

Logarithms (or more usually logs) were invented to make manipulation of awkward numbers easy, but calculators have made it even easier. We can use logs as a tool without a discussion of the theory behind them, but a simple explanation is given at the end of the unit.

Logs are easy on a calculator. You simply press the 'log' key for $\log_{10}$. (Some calculators require that you press the 'log' button before the number and the 'equals' button after the number, others just require the number followed by the 'log' button; practise with your own calculator.)

Inverse logs (or anti-logs) convert a log back into a normal number, and usually involve using two keys, a 'shift' or 'inverse' or 'mode' key and the appropriate 'log' key.

Practise will tell you which key is required for your particular calculator.

Example 1:

	Key sequence
Simple scientific calculator: $\log 6.1 \times 10^{-5}$	
• Numerals: press keys for 6.1	[6], [.], [1]
• One press on the exponential key	[EXP]
• Numeral: 5 (this means '10^5')	[5]
• One press on the $+/-$ key	[+/−]
• One press on the 'LOG' key gives the answer.	[LOG]
• The calculator display will show **−4.21**	

Convert this answer back to its original form using the inverse log key sequence. If you don't get 6.1×10^{-5}, then you have gone wrong somewhere.

Example 2:

	Key sequence
Simple scientific calculator: $\log 4.7 \times 10^{-2}$	
• Numerals 4.7	[4], [.], [7]
• EXP key	[EXP]
• Numeral 2	[2]
• $+/-$ key	[+/−]
• log key	[LOG]
• Answer $= $ **−1.33**	

Example 3:

	Key sequence
Key sequence for a graphical calculator: $\log 6.1 \times 10^{-5}$	
• Press the 'LOG' key	[LOG]
• Numerals, press keys for 6.1	[6], [.], [1]
• One press on the exponential key	[EXP]
• Numeral 5 (this means '10^5')	[5]
• One press on the $+/-$ key	[+/−]
• Press the 'equals' key	[=]
• The calculator will show **−4.21**	

Follow the instructions for your own calculator.

SIGNIFICANT FIGURES, DECIMAL PLACES, AND 'ROUNDING UP'

The calculator will give you an answer correct to six or eight decimal places. In general, only one or two of the numbers after the decimal point are significant. Judge how many significant figures need to be put in the answer by looking at how precise the information was in the original question.

$$e.g. \text{ No. of moles} = \frac{\text{Mass in grammes}}{\text{Relative molecular mass}} \qquad (4.1.1)$$

$$= \frac{6.3}{58} = 0.108\,620\,689$$

But the original mass was only given to two significant figures, so a more sensible answer is 0.11 (rounding the other decimal places up).

Decimal places are simply the series of numbers that occur after the decimal point, *e.g.* 0.246 789 324 2 is expressed to ten decimal places. Significant figures are the important numerals of any number, whether these occur before or after the decimal point. If you are required to give an answer to *three decimal places*, then the answer will have three decimal places, with the others being 'rounded up' as necessary, *e.g.* 36.739 and 0.034 both have three decimal places. If the question requires an answer to *three significant figures*, then the whole answer will only contain three figures (not counting any zeros before the numbers start); 36.7 and 0.000 343 both have three significant figures.

'Rounding up'

Numbers are 'rounded up' by looking at the next figure after the last significant number. If that is above 5, then increase the last significant number by 1. If the number is below 5, then leave the last significant number as it is.

e.g. 3.123 798 to 3 significant figures = 3.12 (3, the third decimal number, is less than 5, so ignore it)
3.127 89 to 3 significant figures (7 is greater than 5, so round up the 2) = 3.13

If you have a number such as 0.003 67, then it will not be possible to give an answer to 2 decimal places, as the final answer would be zero – not a sensible answer. Always try to look at the answer you have from a commonsense point of view and don't just write down the answer the calculator gives you. In any exam requiring calculation, you will invariably lose a mark by writing an answer that gives an absurd number of 'significant' figures . . .

4.1.5 Rearranging Equations

The equations you will normally be given in chemistry will only have one unknown. However, the unknown variable may be embedded in a complex formula. To calculate the unknown, it may be necessary to rearrange the equation.

There are two basic rules to rearranging equations.

TREAT BOTH SIDES OF THE EQUATION THE SAME WAY

As long as you do the same thing to both sides of an equals sign, then that equals sign still holds good. This includes adding or subtracting a number, multiplying, dividing, changing the sign or taking logs of both sides. In the following examples, a few simple equations will be rearranged and solved.

Example 1:
a) If
$$X = 3Y$$

Then we can divide both sides by 3: $\dfrac{X}{3} = \dfrac{3Y}{3}$ giving $Y = \dfrac{X}{3}$ or $X = 3Y$

b) Subtract $3Y$ from both sides:
$$X - 3Y = 3Y - 3Y$$
$$\text{giving } X - 3Y = 0$$

All of these equations are basically the same.

Let us now look at this with a common equation used in chemistry, the relationship between moles, mass in grammes and relative atomic or relative molecular mass. We have already seen this equation using molecular mass on the previous page and will use this equation in Unit 4.3.

$$\text{No. of moles} = \frac{\text{Mass in grammes}}{\text{Relative atomic mass}} \qquad (4.1.1)$$

To calculate the mass in grammes when given the number of moles, the equation can be rearranged by multiplying *both sides* by the RAM and cancelling (see later) to:

No. of moles $\times$ Relative atomic mass

$$= \frac{\text{Mass in grammes} \times \cancel{\text{Relative atomic mass}}}{\cancel{\text{Relative atomic mass}}}$$

so No. of moles $\times$ Relative atomic mass = Mass in grammes

Example 2:
Calcium has a relative atomic mass of 40; how many moles are there in 10 g of calcium?

$$\text{No. of moles} = \frac{\text{Mass in grammes}}{\text{Relative atomic mass}}$$

$$= \frac{10}{40} = 0.25$$

Example 3:
We need 0.35 moles of calcium for a reaction, how much do we need to weigh out?

$$\text{No. of moles} = \frac{\text{Mass in grammes}}{\text{Relative atomic mass}}$$

$$0.35 \text{ moles} = \frac{\text{Mass in grammes}}{40}$$

Rearrange the equation as above,

$$0.35 \text{ moles} \times 40 = \frac{\text{Mass in grammes} \times 40}{40} = 14 \text{ g}$$

Mass = 14 g

CANCELLING OUT

If the same variable appears above and below a division line, the two variables cancel each other out. Or, if there are two negative signs, these also cancel each other out.

Example:
You are asked to calculate X from the equation $G = -RTX$

Starting with the equation: $\quad G = -RTX$

Divide both sides by RT: $\quad \dfrac{G}{RT} \quad \dfrac{-\cancel{RT}X}{\cancel{RT}}$

Cancel RT/RT on the RHS: $\quad \dfrac{G}{RT} = -X$

Make both sides negative: $\quad \dfrac{-G}{RT} = -(-X)$

The two negative signs on the right-hand side (RHS) cancel each other out.

$$\frac{-G}{RT} = X$$

$$\therefore X = \frac{-G}{RT}$$

4.1.6 Graphs

Graphs are a visual way of showing the relationship between two variables.

PLOTTING A GRAPH

* Draw axes at right angles to each other:
    the horizontal is the *x axis*;
    the vertical is the *y axis*.
* Label the axes with the two variables (and their units), and put the correct scale on each axis. The correct scale will be decided from the range of points you have to plot, but will always be uniform. So if you have five points 0.5, 1.2, 1.9, 3.6, 4.9, then the scale will be from 1–5, evenly spread. Don't plot five points with equal distances between them.
* Plot the points and join them up with a smooth line.
* Which variable goes on which axis? When carrying out an experiment, one set of measurements will be pre-set, either by you or the script; *e.g.* 'take readings at 1 minute intervals', or 'add 10 cm³ portions of solution'. This variable will go on the *x* axis. If you are *given* two sets of data to plot, then there are no fixed rules, but time usually goes on the *x* axis.
* Always try to make the graph and its points fill the whole page. Don't try to squeeze more than one graph on one page. This will allow you to spread the points out as much as possible and make any information that you take from the graph more accurate.

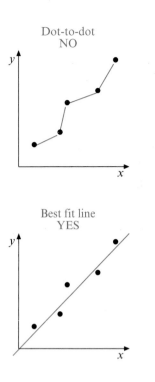

Dot-to-dot
NO

Best fit line
YES

DRAWING THE LINE – THE BEST FIT

When you have plotted the points, you have to join them up with some sort of line. In simple systems, the usual graph shapes are a straight line or a smooth curve. If the graph is a straight line graph, then use a ruler to draw it. In a perfect world the line would go through all the points, but this is rarely a perfect world and it will usually be necessary to draw a line which goes as close as possible to as many points as possible. This is called '*the line of best fit*'. Resist the temptation to draw a separate straight line from one dot to the next. This will give a 'join-the-dots' picture look to the graph, and will not give any impression of a trend in the graph.

If the graph is a curve, then the line joining the points will have to be drawn freehand, but it should still be drawn as a smooth curve which fits the available points as well as possible.

WARNING: Using a best fit line on the graph implies that you actually know what shape the graph should be before you start! The instructions on drawing the graph or the questions that follow will often give you hints about the shape. For instance, you may be told to draw a smooth curve (obviously this graph won't be a straight line) or you may be asked to calculate a gradient (in which case the graph almost certainly will be a straight line).

These days, many people draw graphs by computer. Some computer packages are very sophisticated and will draw 'a best fit' line. The more basic package will still 'join the dots'. If you only have the basic package, it is probably best to draw the line manually.

THE GRADIENT OR SLOPE

The gradient of a straight line graph is given by:

$$\text{Slope} = \frac{y}{x} \text{ or } \frac{\text{vertical}}{\text{horizontal}} \quad \text{(see Figure 4.1.1)}$$

for a graph that passes through the intersection of the x and y axes ('the origin'); and:

$$\text{Slope} = \frac{y_2 - y_1}{x_2 - x_1} \quad \text{(see Figure 4.1.2)}$$

for a graph that doesn't pass through the origin.

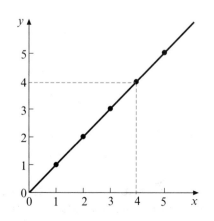

Figure 4.1.1

Graph passes through the origin

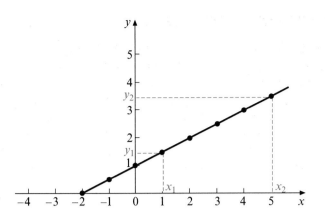

Figure 4.1.2
Graph does not pass through
the origin

Let us try a simple gradient calculation. First the gradient from Figure 4.1.1. At the point on the graph at which $x = 4$, y also $= 4$. The gradient is thus given by $x/y = 4/4$. The gradient is 1.

For Figure 4.1.2, select any two points on the x axis. Let us choose $x = 1$ and $x = 5$. Now draw a horizontal line from those points to the place where they meet the y axis. $x = 1$ corresponds to $y = 1.5$ and $x = 5$ corresponds to $y = 3.5$ The slope from Figure 4.1.2 is given by:

$$\text{Slope} = \frac{y_2 - y_1}{x_2 - x_1} = \frac{3.5 - 1.5}{5 - 1} = \frac{2}{4}$$

The gradient of the graph is 0.5.

USING THE GRAPH

When a graph has been drawn, it can be used to find pieces of information other than the slope. For example, from Figure 4.1.1, we can say what the value of y is at any particular value of x. So at $x = 4$, $y = 4$.

e.g. Using Figure 4.1.2, if $x = 5$, what is the value of y?
   Answer: $y = 3.5$

THE EXPONENTIAL GRAPH

If the graph is not a straight line and looks like Figure 4.1.3 (that is the graph rises progressively more steeply as the x values increase), then it is probably an exponential graph. It is more difficult to measure the slope of a curve, but we can convert the graph to a straight line by doing a 'semi-log' plot.

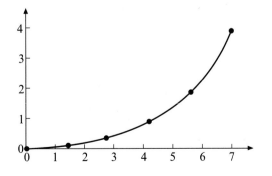

Figure 4.1.3
An exponential graph

Take the exponential logarithm ($\ln_e$) of one set of the values, and plot them against the other set of values without alteration. Exponential logs are taken in a way similar to $\log_{10}$, except using the 'ln' key.

The result will be a straight line if the original plot was exponential. Note that when deciding which set of values to take $\ln_e$ of, it will almost certainly not be time.

These exponential curves are frequently found in experiments on kinetics, measuring the speed of reactions. Here is an example. (To understand the chemistry behind these figures, see Module 5; we are just using a real life example to practise plotting a graph.)

A radioactive isotope of phosphorus ^{31}P has a half-life of 14 days (that means it takes 14 days for the amount of ^{31}P to fall to half its original value, and another 14 days to fall to a quarter, and so on). We can plot a graph of amount of ^{31}P against time in days from these figures;

Relative amount of ^{31}P	40	20	10	5	2.5	1.25
Time (days)	0	14	28	42	56	70

The results are shown in Figure 4.1.4.

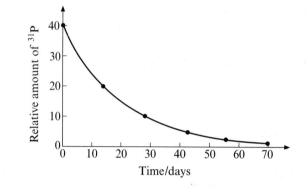

Figure 4.1.4
Exponential decay of ^{31}P

To convert this to a straight line, take the $\ln_e$ of the amounts of ^{31}P remaining. To find the values of $\ln_e$, follow the instructions as above, but using the '$\ln_e$' button. This gives these results,

$\ln_e$(*Relative amount of ^{31}P*)	3.69	3.00	2.30	1.61	0.92	0.22
Time (days)	0	14	28	42	56	70

which are shown in Figure 4.1.5.

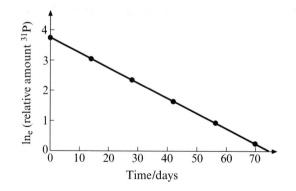

Figure 4.1.5
Semi-log plot of ^{31}P decay

4.1.7 Units

When you were doing maths at school and came to the end of a problem, you were expected to give the correct units for your result. If the units were omitted, the teacher would probably write, 'What are these – elephants?' or a similar sarcastic remark. Units are just as important in chemical calculations as they were in those old maths problems. The particular units for any particular chemical problem will be given in the appropriate section. This is just a guide to help keep the units in order. The units that are used in chemical calculations often seem quite complicated. This is because each component of a particular equation brings with it its own units. To avoid confusion with units, when you are writing out an equation for a calculation and you substitute the figures initially, write the units for each component number down as well. When they are in an equation, they can be multiplied or divided just like the numbers in the equation.

Example 1:

$$\text{Density} = \frac{\text{Mass}}{\text{Volume}}$$

Mass in chemical calculations is usually measured in grammes, g, and volume in cubic centimetres, cm^3.

The units of density are thus g divided by cm^3 (g/cm^3) or g times cm^{-3} ($g\ cm^{-3}$).

The minus sign before a unit, as here with cm^{-3}, tells you that the component is to be divided.

Example 2:
Percentage yield (see Section 4.3.3) is given by,

$$\text{Percentage yield} = \frac{\text{Actual yield}}{\text{Theoretical yield}} \times 100$$

This can be expressed in either 'mass in g' or 'moles', so:

$$\text{percentage yield} = \frac{\text{actual yield (g)}}{\text{theoretical yield (g)}} \times 100 \quad \text{or}$$

$$\frac{\text{actual yield (moles)}}{\text{theoretical yield (moles)}} \times 100$$

Because the units cancel out in each case, the final answer is the same from both types of calculation.

Answers to diagnostic test

If you score less than 80%, then work through the unit and re-test
yourself at the end using this same test. If you still get a low score,
then re-work the unit at a later date.

1 $Y = \dfrac{6X - 4}{3}$ (2)

2 3.97×10^{-6} (2)

3 -5.401 (2)

4 3.87 (2)

5 6×10^{11} (2)

10 Marks   (80% = 8)

FURTHER QUESTIONS (ANSWERS GIVEN IN APPENDIX)

Write the following as ordinary numbers.

1 9.663×10^5

2 7.28×10^1

3 4.82×10^{-3}

Write the following in standard form.

4 $2\,020\,000$

5 $0.000\,001\,11$

6 2840

Correct the following numbers to three decimal places.

7 $0.007\,69$

8 $0.008\,8$

9 3.223

Correct the following numbers to 3 significant figures.

10 $0.007\,69$

11 $0.008\,8$

12 3.223

13–21 Take the logarithms to base 10 ($\log_{10}$) of the numbers in
Questions 1–9.

Perform the following calculations without using a calculator.

22 $(7 \times 10^5) + (5.001 \times 10^5)$

23 $(7.28 \times 10^3) + (9.43 \times 10^4)$

24 $(9.91 \times 10^6) \times (1.5 \times 10^5)$

25 $(4.00 \times 10^3) \times (2.7 \times 10^6)$

26 Rearrange the equation below to find C.

$A = B + nC$

27 Rearrange the equation below to find X.

$C = \dfrac{BAX}{50}$

28 The speed of a car at various times is given by the table below.
Plot a graph of speed (vertical or y axis) against time (horizontal

or x axis). From the graph answer the following questions.
a) What is the speed of the car after 10 sec?
b) How long does it take for the car to achieve a speed of 60 km h^{-1}?
c) What is the acceleration of the car? (The acceleration of the car is the rate of change of speed with time, in km h^{-1} s^{-1}.) This can be found by measuring the gradient of the graph.

Speed (km h^{-1})	0	32	64	96	128
Time (s)	0	4	8	12	16

4.1.8 Extension – Theory of Logarithms

Please make sure that you have worked through the section on conversion of ordinary numbers to standard form before reading this.

LOGARITHMS TO BASE 10 (Abbreviated to log$_{10}$ or, simply, log)

We have said that a number in standard form has the form:

$$A \times 10^n$$

(*A* must always be between 1 and 10)
(*n*, the number to which ten is raised, is the number of places the decimal point 'moves')

The number to which the '10' has to be raised in the examples we have looked at has always been a whole number, *e.g.* 10^6 or 10^{-5}.

In logs, a logarithm to base 10 of *any* number is the power to which 10 has to be raised in order to equal that number.

Take a simple example, $1000 = 10^3$. Therefore, the log of $1000 = 3$. Similarly, the log of 1 million $(1\,000\,000 = 10^6) = 6$.

So to work out the log of a number which is a multiple of 10, you don't even need a calculator, simply convert it to standard form and the number to which the 10 has to be raised equals the log.

Example 1:
Try this: What is the log of $100\,000\,000\,000\,000\,000\,000\,000$?
Answer: Follow the instructions for converting to standard form in Section 4.1.2. The number has 23 zeros, so this equals 10^{23}, therefore the log $= 23$.

Of course, most numbers are not straight multiples of 10. This will result in powers of 10 that are *not* whole numbers, *e.g.* $10^{3.316}$. For these it is easiest to use a calculator, using the 'log' button. Let us try one example (the key sequence is not given here, see Unit 4.1 for actual instructions).

Example 2:
What is the log$_{10}$ of 1584? This means that we need to know the number to which 10 has to be raised give that number.

The calculator tells us that log$_{10}$ 1584 $= 3.1998$. So, $10^{3.1998} = 1584$. Check this out for yourself.

Notice that if we convert 1584 to standard form, it becomes 1.584×10^3, and the log has the number 3 as its whole number – this is because the log of 10^3 is 3. The log can thus give you an idea of the relative size of a number without even doing an exact calculation. Of course, we don't bother converting numbers to standard form just to get the log. We just use the log button on the calculator.

What about reverse calculations, known as inverse logs?

Example 3:
Take the $\log_{10} = 3.316$. This equals the number $10^{3.316}$. The calculator tells us that the inverse log or the number that has this log is 2070.1.
So $\log_{10} 2070.1 = 3.316$.

LOGARITHMS TO THE BASE e (Natural logs, abbreviated to $\ln_e$)

The symbol 'e' is a mathematical constant (e = 2.718 . . .).

The logarithm to base e of a number is the power to which the 'e' has to be raised to get that number. For these calculations we use the 'ln' button on the calculator.

So, $\ln_e 51963 = 10.8583$ (= $e^{10.8583}$).

Unit 4.2
Chemical Laws – The Law of Constant Composition and the Law of Conservation of Mass

Starter check

What you need to know before starting the contents of this unit.

You should already know (Sections 1.3.3 and 1.3.4) how to calculate atomic masses from the numbers of protons and neutrons in the nucleus, and that isotopes are taken into account when determining the relative atomic masses of the elements.

Aims **By the end of this unit you should understand:**

- That the formula of a compound is fixed and unchanging.
- That in a chemical reaction, mass can neither be created nor destroyed.
- How to balance a chemical equation.

Diagnostic test

Try this test at the start of the unit. If you score more than 80%, then use this unit as a revision for yourself and scan through the text. If you score less than 80%, then work through the text and re-test yourself at the end using this same test.

The answers are at the end of the unit.

1 If the relative atomic masses of calcium, carbon and oxygen are 40, 12 and 16, respectively, what is the relative formula mass of calcium carbonate, $CaCO_3$? **(2)**

2 If 10 g of calcium carbonate decomposed on heating to calcium oxide and carbon dioxide, what total mass of calcium oxide and carbon dioxide would be produced?

$$CaCO_3(s) \xrightarrow{\text{heat}} CaO(s) + CO_2(g)$$ **(3)**

3 Balance the following equation.

$$Fe(s) + HCl(aq) \rightarrow FeCl_2(aq) + H_2(g)$$ **(2)**

4 State the law of constant composition. (2)

5 Write out and balance the equations for the following chemical reactions:

a) Calcium carbonate + hydrochloric acid to give calcium chloride, carbon dioxide and water. (2)

b) Aluminium powder + sulfur to make aluminium sulfide. (2)

c) Sulfuric acid + potassium hydroxide to make potassium sulfate and water. (2)

15 Marks   (80% = 12)

4.2.1 The Law of Constant Composition

If we consider water, H_2O, it always contains hydrogen and oxygen in the proportion, two atoms of hydrogen to one of oxygen. Whether the water comes from the melting of pure ice or the condensing of steam, which are physical processes, or whether it is produced in chemical reactions such as the burning of hydrogen or the neutralisation of an alkali by an acid makes no difference – all of the water has the same composition.

$$2H_2(g) + O_2(g) \rightarrow 2H_2O(l)$$

$$NaOH(aq) + HCl(aq) \rightarrow NaCl(aq) + H_2O(l$$

This idea – that once a chemical formula for a compound is established, that compound always has the same formula – is expressed in the law of constant composition:

A compound always contains the same elements in the same proportions by mass, no matter how that compound is prepared.

Constant mass proportions were originally determined by analysing the compound copper carbonate, $CuCO_3$, in the 1790s. We now know that no matter how the copper carbonate is prepared it always contained atoms of copper, carbon, and oxygen in the proportions 1:1:3.

Any scientific law summarizes the observations made in experiments by very many people. A law is only called a Law when, whenever it is tested by experiment, it is found to hold true. (It is always a big event in science when someone produces an experimental result, which can be repeated by many other scientists, which contradicts a Law! The Law then has to be scrapped and replaced by something better.)

4.2.2 The Law of Conservation of Mass

A person carrying a heavy load can often be heard to say, 'The longer I carry this parcel, the heavier it gets!' This is not true, of course, the person's arms are simply starting to ache. The mass of the parcel remains the same. We cannot increase the mass of something over time without loading its contents further. Similarly the story of 'the magic porridge pot', which continues to produce food even after it is shown to be empty, is simply a myth. It just isn't possible to produce something from nothing! It also follows that we cannot get rid of something either – the disappearing lady in a conjuring trick is merely an illusion. This theory is expressed scientifically in the law of conservation of mass:

The total mass of the reacting substances before a chemical reaction is always equal to the mass of the substances produced by the reaction.

We usually simplify this to:

Mass can neither be created nor destroyed in a chemical reaction.

To see how this works, let us consider the formation of carbon dioxide, which is produced by combining carbon and oxygen in the proportion 1:2.

In Section 1.3.1, we learnt that the mass of an atom is effectively in its nucleus, and that if we add together the number of protons and neutrons in the nucleus we get the mass number. We can therefore say that the mass of an atom in atomic mass units (a.m.u.) is equal to the sum of the number of protons and neutrons in its nucleus. Carbon, with 6 protons and 6 neutrons has a mass number of 12, and oxygen with 8 protons and 8 neutrons has a mass number of 16. A molecule of oxygen must have a mass of 32 mass units, since it contains two oxygen atoms.

The law of conservation of mass tells us that the carbon dioxide that is formed by burning carbon will have a mass of 44 mass units, made up of 6 protons and 6 neutrons from the carbon and 8 protons and 8 neutrons from each of the oxygens $[(6 + 6 + 8 + 8 + 8 + 8) = 44$; or $12 + (2 \times 16) = 44]$.

$$C(s) \quad + \quad O_2(g) \quad \rightarrow \quad CO_2(g)$$

6 protons	2×8 protons	$(6 + 8 + 8)$ protons
+ 6 neutrons	+ 8 neutrons	+ $(6 + 8 + 8)$ neutrons
12	+ (2×16) $\rightarrow$	44 (in a.m.u.)

In the next unit we will see how these 'atomic mass units' relate to masses that we can actually measure, but for now we can just consider the mass in terms of total numbers of protons and neutrons.

Let us consider again the formation of water, which is produced by combining hydrogen and oxygen in the proportions 2:1, or in terms of atomic mass units, 2:16 or 1:8. One atomic mass unit of hydrogen will react with 8 atomic mass units of oxygen to produce 9 a.m.u. of water. Also, if we break the water down to hydrogen and oxygen, by passing an electric current through it, we know how much hydrogen and oxygen to expect. If we start with 180 a.m.u. of water, then we will get 20 a.m.u. of hydrogen and 160 a.m.u. of oxygen. These are always in the same proportion of 1:8 by mass.

$$2H_2(g) \quad + \quad O_2(g) \quad \rightarrow \quad 2H_2O(l)$$
$$(2 \times 2) \text{ a.m.u.} \quad 32 \text{ a.m.u.} \quad (2 \times 18) \text{ a.m.u.}$$

or
$$H_2(g) \quad + \quad \frac{1}{2}O_2(g) \quad \rightarrow \quad \frac{1}{2}H_2O(l)$$
$$2 \text{ a.m.u.} \quad 16 \text{ a.m.u.} \quad 18 \text{ a.m.u.}$$

Sometimes it isn't obvious that there has been no reduction in mass. For example, in a camping gas burner the propane (C_3H_8) undergoes combustion to carbon dioxide and water, both of which are lost to the atmosphere. It is important to remember that these compounds are merely lost to the reaction vessel and are still out there somewhere! In this example, carbon has a relative atomic mass of 12, hydrogen 1 and oxygen 16; one molecule of propane thus has a mass of $(3 \times 12) + 8 = 44$ a.m.u. and 5 molecules of oxygen $5 \times 32 = 160$ a.m.u.

$$C_3H_8(g) + 5O_2(g) \rightarrow 3CO_2(g) \quad + \quad 4H_2O(l)$$

	44	160	$3 \times (12 + 32)$	$4 \times (2 + 16)$
			= 132	= 72
Total	204 a.m.u.		204 a.m.u.	

4.2.3 Balancing Equations

When we write down a chemical reaction in the form of an equation, we have to balance the equation to ensure that we don't make mass appear or disappear. That is to say, the number of atoms of any particular element that are present at the beginning of the reaction, on the left-hand side or LHS of the arrow, is equal to the number on the right-hand side (RHS). Looking at a simple equation such as the combustion or burning of hydrogen:

$H_2(g) + O_2(g) \rightarrow H_2O(l)$
Left-hand side (LHS)   Right-hand side (RHS)

The molecule of water always contains 1 oxygen atom, and the molecule of oxygen always contains 2 atoms of oxygen. The above equation has thus managed to lose an atom of oxygen in going from the LHS to the RHS. The law of conservation of mass does not allow this! In order to correct this we must 'double up' the number of molecules of water:

$H_2(g) + O_2(g) \rightarrow \mathbf{2} \times H_2O(l)$
LHS   RHS

This now gives 2 atoms of oxygen on both sides of the equation. But, now there are 4 atoms of hydrogen on the RHS of the equation and only 2 on the left. To even this out, 'double up' the number of hydrogen atoms:

$\mathbf{2} \times H_2(g) + O_2(g) \rightarrow \mathbf{2} \times H_2O(l)$
Left-hand side (LHS)   Right-hand side (RHS)

It is easier to see this if we set out the data as in the following table:

Element	Before balancing LHS	Before balancing RHS	After balancing LHS	After balancing RHS
Hydrogen	2	2	4	4
Oxygen	2	1	2	2

After balancing, the number of atoms on the LHS should be the same as the number on the RHS.

Another example is the combustion of propane. So that we can concentrate on the balancing, the states (g), (l), *etc.* have been omitted.

$C_3H_8 + O_2 \rightarrow CO_2 + H_2O$ unbalanced becomes

$C_3H_8 + 5O_2 \rightarrow 3CO_2 + 4H_2O$
        LHS   RHS

Element	Before balancing LHS	Before balancing RHS	After balancing LHS	After balancing RHS
Carbon	3	1	3	3
Hydrogen	8	2	8	8
Oxygen	2	3	10	10

Let us now check for all the elements in the equation,

Element	LHS	RHS
Na	2	2
O	6	6
S	1	1
H	4	4

We have the same number of atoms of each of the elements on each side of the equation, so the equation is balanced.

Example 2:

$$Al(s) + FeO(s) \rightarrow Al_2O_3(s) + Fe(s)$$

The compounds have the correct formulae. Let us start with aluminium. Initially:

Element	LHS	RHS
Al	1	2

Double up the amount of aluminium on the LHS and put it in the table.

Element	LHS	RHS
Al	2	2

The equation becomes: $2Al + FeO \rightarrow Al_2O_3 + Fe$

Now consider the effect on the oxygen.

Element	LHS	RHS
O	1	3

We must introduce a '3' in front of the FeO on the LHS to correct the equation.

$$2Al + 3FeO \rightarrow Al_2O_3 + Fe$$

Element	LHS	RHS
Al	2	2
O	3	3

OK, so far, but what effect has this had on the iron?

Element	LHS	RHS
Fe	3	1

We must introduce a '3' in front of the Fe on the RHS:

$$2Al + 3FeO \rightarrow Al_2O_3 + 3Fe$$

Element	LHS	RHS
Al	2	2
O	3	3
Fe	3	3

The equation is now balanced.

Example 3:

$$CaCO_3(s) + H_3PO_4(aq) \rightarrow Ca_3(PO_4)_2(s) + CO_2(g) + H_2O(l)$$

This looks on the surface quite a complicated reaction, but let us look more closely and logically. In the first example we consider the components of 'SO_4' as separate elements, but we could have considered them as a group. In this case, the group 'PO_4' occurs on both sides of the equation, so we can consider that as one unit (see Hint 2).
    Start with a table of elements.

Element	LHS	RHS	
Ca	1	3	
PO₄	1	2	
C	1	1	
O	3	3	(don't count the oxygen atoms in PO_4 units)
H	3	2	

Let's start with calcium. We need three of those on the LHS.

$$3CaCO_3(s) + H_3PO_4(aq) \rightarrow Ca_3(PO_4)_2(s) + CO_2(g) + H_2O(l)$$

Element	LHS	RHS	
Ca	3	3	
PO₄	1	2	
C	3	1	
O	9	3	(don't count the oxygen atoms in PO_4 units)
H	3	2	

We are also short of a 'PO_4' on the LHS, so let us double up the H_3PO_4.

$$3CaCO_3(s) + 2H_3PO_4(aq) \rightarrow Ca_3(PO_4)_2(s) + CO_2(g) + H_2O(l)$$

Element	LHS	RHS	
Ca	3	3	
PO₄	2	2	
C	3	1	
O	9	3	(don't count the oxygen atoms in PO_4 units)
H	6	2	

Whoops! Has that made things worse? The hydrogens are not balanced, and nor are oxygens and carbons. But if we look carefully we can see that putting a '3' in front of any carbon-, hydrogen- or oxygen-containing molecule on the RHS will sort that out.

$$3CaCO_3(s) + 2H_3PO_4(aq) \rightarrow Ca_3(PO_4)_2(s) + 3CO_2(g) + 3H_2O(l)$$

Element	LHS	RHS
Ca	3	3
PO$_4$	2	2
C	3	3
O	9	9
H	6	6

(don't count the oxygen atoms in PO$_4$ units)

Balanced!

And now that it is, we can reveal that this reaction DOES NOT HAPPEN! – calciumn phosphate is insoluble (which is just as well, since large amounts of it are in your bones and teeth). As soon as any particles of calcium carbonate react with phosphoric acid, the reaction stops because the particles are coated with insoluble calcium phosphate. So now we can write Rule 8 (see start of Section 4.2.4 on page 142):

8 MAKE SURE THAT YOUR EQUATION IS FOR A REACTION THAT ACTUALLY HAPPENS.

This description makes the process of balancing an equation seem very long-winded, but as you get used to doing more of them, it gets faster. Now try the supplementary questions on this topic for extra practice. You'll find them after the answers to the diagnostic test. Some of them may not need changing in order to be balanced – look at each example logically and add up the number of atoms on each side first. Don't just jump to conclusions!

Answers to diagnostic test

If you score less than 80%, then work through the text and re-test yourself at the end using this same test. If you still get a low score then re-work the unit at a later date.

1 The relative formula mass of calcium carbonate, $CaCO_3$ is 100. **(2)**

2 The total mass of calcium oxide and carbon dioxide would be same as on the right-hand side, 10 g – made up of 5.6 g of CaO and 4.4 g of CO_2. **(3)**

3 The balanced equation is:

$$Fe(s) + 2HCl(aq) \rightarrow FeCl_2(aq) + H_2(g)$$ **(2)**

4 A compound always contains the same elements in the same proportions by mass, no matter how that compound is prepared. **(2)**

5 The balanced equations are:
  a) $CaCO_3(s) + 2HCl(aq) \rightarrow CaCl_2(s) + CO_2(g) + H_2O(l)$ **(2)**
  b) $2Al(s) + 3S(s) \rightarrow Al_2S_3(s)$ **(2)**
  c) $H_2SO_4(aq) + 2KOH(aq) \rightarrow K_2SO_4(aq) + 2H_2O(l)$ **(2)**

15 Marks   (80% = 12)

FURTHER QUESTIONS (ANSWERS GIVEN IN APPENDIX)

Any atomic masses can be looked up on the periodic table.

1 Glucose isolated from grape sugar contains the elements carbon, hydrogen and oxygen in the ratio of 6:1:8 by mass. What is the formula of glucose isolated from blood sugar?

2 A camping gas canister contains 500 g of propane. When it has been completely burnt, 1500 g of carbon dioxide and 818 g of water are produced. Calculate the mass of oxygen that has been used up in the combustion reaction.

3 Balance the following equations:

  a) Hydrogen gas reacts with chlorine gas to give hydrogen chloride.

$$H_2(g) + Cl_2(g) \rightarrow HCl(g)$$

  b) Heating barium carbonate causes it to decompose to solid barium oxide and carbon dioxide.

$$BaCO_3(s) \rightarrow BaO(s) + CO_2(g)$$

  c) A solution of sodium chloride reacts with a solution of silver nitrate to give a solution of sodium nitrate and solid silver chloride.

$$NaCl(aq) + AgNO_3(aq) \rightarrow NaNO_3(aq) + AgCl(s)$$

  d) Combustion of aluminium giving aluminium oxide.

$$Al(s) + O_2(g) \rightarrow Al_2O_3(s)$$

e) The neutralisation of calcium hydroxide with nitric acid.

$$Ca(OH)_2(s) + HNO_3(aq) \rightarrow H_2O(l) + Ca(NO_3)_2(aq)$$

For Questions 4–6, refer back to Unit 3.1 for the different types of reaction if you aren't sure of them.

4 Write and balance an equation for the reaction of manganese(II) sulfate and sodium hydroxide in aqueous solution.

5 Write and balance an equation for the combustion of ethanol (C_2H_5OH).

6 Complete and balance the neutralisation reaction.

$$HCl(aq) + Mg(OH)_2(s) \rightarrow$$

Unit 4.3
Reacting Quantities and the Mole

Starter check

What you need to know before starting this unit.

You need to be familiar with the use of exponential numbers in calculations and how to use your calculator. You should know how to calculate relative formula masses and be familiar with isotopes and the meaning of isotopic composition. You should also know how to balance equations.

Aims **By the end of this unit you should understand how to:**

- Define the mole.
- Calculate the number of moles of a substance from its mass in grammes.
- Use moles to calculate the amount of product that will be formed in a chemical reaction in either moles or grammes, even if one of the reagents is present in excess.
- Calculate a percentage yield.
- Convert percentage composition into an empirical formula.

Diagnostic test

Try this test at the start of the unit. If you score more than 80%, then use this unit as a revision for yourself and scan through the text. If you score less than 80%, then work through the text and re-test yourself at the end using this same test.

The answers are at the end of the unit.

Relative atomic masses: silver, 108; nitrogen, 14; oxygen, 16; carbon, 12; chlorine, 35.5; calcium, 40; sodium, 23.

1 Using the periodic table, what is the relative atomic mass of the element xenon? **(1)**

2 What is the relative formula mass of the compound sodium chloride? **(1)**

3 How many grammes of calcium are present in 25 g of calcium carbonate? **(4)**

4 You are asked to weigh out 0.1 moles of calcium carbonate. What mass would you weigh out? **(4)**

5 What mass of sodium chloride must be used to complete the following reaction, if we start with 1.7 g of silver nitrate?   **(4)**

$NaCl(aq) + AgNO_3(aq) \rightarrow NaNO_3(aq) + AgCl(s)$

6 The amount of calcium carbonate formed in a reaction was expected to be 8.7 g, but only 6.9 g was obtained. What percentage yield is that?   **(2)**

7 The percentage composition of glucose is 40% carbon, 6.7% hydrogen, and 53.3% oxygen. What is its empirical (or simplest ratio of atoms) formula?   **(4)**

20 Marks   (80% = 16)

4.3.1 Introduction to the Mole

You probably first heard this trick question in primary school.

    Q: What weighs more, a pound of feathers or a pound of lead?
    A: Silly question! They both weigh the same, of course.

These days we talk of kilogrammes not pounds, but the same is true of a kilogramme of charcoal, which is just atoms of pure carbon, and a kilogramme of atoms of pure lead. But even though they have the same mass, you don't seem to be getting much lead for your money, when compared with the amount of carbon.

What is the difference? The mass of an atom comes from the number of protons and neutrons present in the nucleus; very little contribution comes from the electrons, as they are so small. Lead has 82 protons and 125 neutrons, making a mass number of 207, whereas carbon has only 6 protons and 6 neutrons, (atomic mass 12) and hydrogen, with only 1 proton has an a mass number of 1. It follows that 207 kilogrammes of lead will contain the same number of atoms of lead as there are carbon atoms in 12 kilogrammes of carbon, and the same number of atoms as there are hydrogen atoms in one kilogramme of hydrogen.

It makes chemical life simpler to talk about the mass of an atom as if it has the mass of the total number of protons and neutrons, and set the mass of a proton and a neutron at 1 unit. This is usually called an atomic mass unit.

What is the actual value of this unit?

Atoms individually are very small: one gramme of carbon contains about 20 thousand million, million, million atoms, so weighing out small numbers of atoms is impossible. But it has been found that if you take *the relative atomic mass* (see Section 1.3.4) in grammes of any element, it will contain the same number of atoms as the relative atomic mass of any other element. So 12 g of carbon will have the same number of atoms of carbon as 207 g of lead has atoms of lead.

The actual number of atoms in the relative atomic mass in grammes of any element is 6×10^{23} and is known as **Avogadro's constant**. Its value was determined in the early years of this century, and confirmed by several independent methods. The amount of any element or compound that contains this number of particles is called 'one mole' of that substance.

The mole is the SI unit for the measurement of chemical substance and its official definition is:

> The size of the number 6×10^{23} can be visualised like this: it is about the number of grains of sand on all the beaches in the world, or the number of red corpuscles in the blood of all the people in the world.

'The mole is that amount of chemical substance that contains as many particles (molecules, formula units, atoms or ions) as there are atoms of ^{12}C in exactly 12.0 g of carbon-12.'

We can summarise this by saying:

In practice, one 'mole' is the relative atomic or molecular or formula mass of any substance expressed in grammes, and it contains 6×10^{23} particles. We do not have to go around counting the particles; instead we can use the relative masses of elements or compounds. These are easily measured by weighing quantities.

If you look at the periodic table on page ii at the front of the book, you can find the relative atomic mass of any element. Some of these RAMs

4.3.4 Molar Excess and Limiting Quantities

Consider the above example of magnesium carbonate reacting with sulfuric acid:

$$MgCO_3(aq) + H_2SO_4(aq) \rightarrow MgSO_4(s) + CO_2(g) + H_2O(l)$$
10 g (RFM = 84)   (RFM = 120)

The acid is a common laboratory reagent, so it is quite likely that, when doing such a reaction, we would weigh out a mass of magnesium carbonate and calculate how much acid was required to do the reaction, and then add a bit more acid, just to make sure that there was enough to complete the reaction. In other words, the sulfuric acid is in excess of the amount that is actually required, and consequently the amount of magnesium sulfate that is produced is limited by the amount of magnesium carbonate we started with and not the sulfuric acid.
    Generalising this:

Any chemical completely used up in the reaction is the **limiting reagent**. The chemical which is left over at the end is the **excess reagent** (usually this is the cheapest or most common chemical!).

It is important to know if one of the reactants is in excess before calculating a percentage yield.

How do we know which reagent is in excess?

Let us take an example. Treatment of iron(II) sulfide with hydrochloric acid is a way of generating hydrogen sulfide. (RAMs: Fe, 55.8; S, 32.1; Cl, 35.5; H, 1, respectively.)

$$FeS(s) + 2HCl(aq) \rightarrow FeCl_2(aq) + H_2S(g)$$
RMM 87.9   36.5   126.8   34.1

If we start with 50 g of iron sulfide and 50 g of hydrogen chloride (in hydrochloric acid), which reagent is in excess? Remember, in this reaction we require 2 moles of hydrochloric acid for each mole of FeS.
    First, calculate how many moles of each reagent we have used. Using the formula for calculating moles; equation. 4.1.1:

No. of moles of FeS $= 50/87.9$ $= 0.57$ moles
No. of moles of HCl needed
to react exactly with the FeS $= 2 \times 0.57 = 1.14$ moles
No. of moles of HCl available $= 50/36.5$ $= 1.36$ moles

So more hydrochloric acid has been added than was actually necessary; the hydrochloric acid is in excess and the iron sulfide is the limiting reagent. We must base the % yield on the number of moles of iron sulfide.
    Each mole of FeS yields 1 mole of H_2S, so the maximum or theoretical yield of H_2S is 0.57 moles (or, as mass = moles $\times$ RMM, mass of H_2S = 0.57 moles $\times$ 34.1 = 19.40 g).

Let us work through another example.
Consider the following reaction:

$$MgSO_4(aq) + BaCl_2(aq) \rightarrow MgCl_2(aq) + BaSO_4(s)$$

First, work out the RMMs and RFMs from the periodic table:

RAMs: S, 32.1; O, 16; Mg, 24.3; Cl, 35.5; Ba, 137.3

	$MgSO_4(aq)$ +	$BaCl_2(aq)$ $\rightarrow$	$MgCl_2(aq)$ +	$BaSO_4(s)$
RMM/RFM	120.4	208.3	95.3	233.4

Next, check that the equation is balanced. It is, so try the following questions.

QUESTIONS

a) If we start with 10 g of $MgSO_4$, how much $BaCl_2$ is needed to react with it exactly?
b) What theoretical yield of $BaSO_4$ should we expect?
c) If we only get 15 g of $BaSO_4$, what is the percentage yield?
d) If we start with 5 g of $MgSO_4$ and 4 g of $BaCl_2$, is one of the reagents in excess, and if so, which one is the limiting reagent?
e) Using the information from question (d), what is the theoretical yield?

ANSWERS

a) 10g of $MgSO_4$ = mass (g)/RFM = 10/120.3 = 0.083 mol, and we need an equal number of moles of $BaCl_2$. Therefore, we need 0.083 mol of $BaCl_2$.
No. of moles × RFM = mass needed, so 0.083 × 208.3 = 17.3 g of $BaCl_2$ is needed.
b) 1 mole of $MgSO_4$ will yield 1 mole of $BaSO_4$, so we can onlya expect to get 0.083 mol from 0.083 mol of $MgSO_4$. As in question (a), moles × RFM = mass in grammes, so:
yield = 0.083 × 233.3 = 19.36 g
c) A yield of 15 g represents a percentage yield of 15/19.36 × 100 = 77.5%.
d) 5g of $MgSO_4$ = 5/120.3 = 0.042 mol, and 4 g of $BaCl_2$ is 4.208.3 = 0.019 mol. Therefore, the $MgSO_4$ is in excess, the $BaCl_2$ is the limiting reagent, and the yield must be based on the amount of $BaCl_2$ (0.019 mol).
e) The maximum yield will be 0.019 mol or 0.019 × 233.3 = 4.43g.

4.3.5 Percentage Composition, and using the Law of Constant Composition to find the Empirical and Chemical Formula of a Compound

When a gas is burnt and the vapour collected, a colourless liquid is produced. We have reason to think that the gas is hydrogen and that the liquid is water. A sample of the liquid is collected and its percentage composition by weight is determined as 11% hydrogen and 89% oxygen. We can then say that 100 g of the liquid contains 89 g of oxygen and 11 g of hydrogen. Since we know that 1 mole of hydrogen atoms has a mass of 1 g and 1 mole of oxygen atoms has a mass of 16 g, then the number of moles of each element present is:

$$\text{No. of moles of O} = \frac{\text{mass (g) of O}}{\text{RAM of O}} = \frac{89 \text{ g}}{16 \text{ g}} = 5.56$$

$$\text{No. of moles of H} = \frac{\text{mass (g) of H}}{\text{RAM of H}} = \frac{11 \text{ g}}{1 \text{ g}} = 11.00$$

This gives a formula of $H_{11}O_{5.6}$.

A molecular formula is expressed in small whole numbers, so we have to reduce this to its lowest common denominator by dividing both numbers by 5.6. That is $H_{1.96}O_1$, which is very close indeed to H_2O. Thus the liquid is confirmed as water.

Example 1:
A compound of iron and bromine contains 25.9% iron and 74.1% bromine. What is its formula?
a) First find the RAMs of iron and bromine from the periodic table. These are 55.8 and 79.9 for Fe and Br, respectively.
b) Then using equation 4.1.1, No. of moles = mass in grammes/RAM, assuming we have 100 g of compound, we can say:

$$\text{No. of moles of Fe} = \frac{\text{mass (g) of Fe}}{\text{RAM of Fe}} = \frac{25.9 \text{ g}}{55.8 \text{ g}} = 0.464$$

$$\text{No. of moles of Br} = \frac{\text{mass (g) of Br}}{\text{RAM of Br}} = \frac{74.1 \text{ g}}{79.9 \text{ g}} = 0.927$$

c) This gives the formula $Fe_{0.464}Br_{0.927}$. We simplify this by noting that 0.464/0.927 is very nearly equal to 1/2, so the formula is $FeBr_2$.

Example 2:
A compound contains 62.6% lead, 8.5% nitrogen and 29.0% oxygen. What is its formula?
a) The RAMs of the elements are lead, 207.2; nitrogen, 14; and oxygen, 16.
Using the same method we get:

$$\text{No. of moles of Pb} = \frac{62.6 \text{ g}}{207.2 \text{ g}} = 0.302$$

$$\text{No. of moles of N} = \frac{8.5 \text{ g}}{14 \text{ g}} = 0.607$$

$$\text{No. of moles of O} = \frac{29.0 \text{ g}}{16 \text{ g}} = 1.813$$

The formula becomes $Pb_{0.3}N_{0.6}O_{1.8}$. Divide each of those numbers by 0.3 (the smallest number) and the formula becomes $Pb_1N_2O_6$. This answer is correct.

However, because we know that nitrates have the formula NO_3, we can group 1 N with 3 O atoms and rewrite this as $Pb(NO_3)_2$; the compound is lead(II) nitrate.

Example 3:
This is a slightly more complex example. A bottle of an amino acid has lost its label. A sample is sent for analysis and the percentage composition is reported as:

C   32.12%   RAM 12
H   6.71%   RAM 1
O   42.58%   RAM 16
N   18.59%   RAM 14

Let us look at carbon first:

$$32.12 \text{ g C} \times \frac{1 \text{ mole of C}}{12 \text{ g C}} = 2.67$$

Similarly, for the other elements:

H   6.71 $\times$ 1/1 = 6.71
O   42.58 $\times$ 1.16 = 2.66
N   18.58 $\times$ 1/14 = 1.33

This gives a composition of C, 2.67; H, 6.71; N, 1.33; O, 2.66. To convert this to small whole numbers, we can divide each of the numbers by 1.33 (the smallest number). That will take the number of nitrogens present to 1. So: C, 2.67/1.33; H, 6.71/1.33; N, 1.33/1.33; O, 2.66/1.33 = $C_{2.00}H_{5.0}NO_{2.0}$ or $C_2H_5NO_2$.

The amino acid glycine, H_2NCH_2COOH, has the formula $C_2H_5NO_2$.

You might find this easier to follow if it is set out as in Table 4.3.1:

Table 4.3.1

Element	C	H	O	N
Mass or %	32.12	6.71	42.58	18.58
RAM	12	1	16	14
No. of moles (mass/RAM)	2.68	6.71	2.66	1.33
Divide by smallest value (1.33)	2.01	5.04	2	1
Atom ratio	2	5	2	1

$\therefore$ Empirical formula is $C_2H_5O_2N$

DIFFERENCE BETWEEN THE CHEMICAL FORMULA AND THE EMPIRICAL FORMULA

When we write out a chemical formula, we put in the exact number of atoms of each of the elements that occur in a molecule, or in the formula of an ionic compound. For example, water is H_2O, calcium carbonate is $CaCO_3$, and glucose is $C_6H_{12}O_6$.

An **empirical formula** shows which elements are in a compound and the simplest ratio of the atoms of those elements. For water and calcium carbonate, the chemical formula does have the smallest possible whole numbers, but for glucose, we can write CH_2O. This is the empirical formula of glucose, it is not the actual molecular formula.

The method of calculating formulae from percentage compositions that we have just been using gives the empirical formula, or relative numbers of atoms, in a molecule. In the examples that we have been looking at, the empirical formula is identical to the chemical formula. Compounds with the molecular formula $C_4H_{10}N_2O_4$ or $C_6H_{15}N_3O_6$ would have the same percentage compositions as glycine. In order to be certain that the formula we have is the molecular formula, we must have an RMM to go with it. So, if we also knew that the relative molecular mass of the compound was 75, we then could be certain that the compound was $C_2H_5NO_2$.

This difference between empirical formula and chemical formula (known for covalent compounds as the *molecular formula*) becomes important when we look at organic chemistry. (See Module 6.)

REVERSE CALCULATIONS; PERCENTAGE COMPOSITION FROM FORMULA

We can turn this sort of calculation around and convert chemical formula into percentage composition.

Example 4:
Let us take calcium carbonate, $CaCO_3$, RFM 100. Calcium has a RAM of 40, so the mass of calcium contributes $^{40}/_{100}$ths to the mass of 1 mole of calcium carbonate.

$$\text{Percentage composition} = \frac{40}{100} \times 100 = 40\%$$

Carbon contributes 12/100 and oxygen $(3 \times 16)/100$; their percentages are 12% and 48%, respectively.

Example 5:
Calculate the percentage composition of glucose, $C_6H_{12}O_6$, RMM 180.

Carbon: $\dfrac{(6 \times 12)}{180} \times 100 = 40\%$

Hydrogen: $\dfrac{(12 \times 1)}{180} \times 100 = 6.7\%$

Oxygen: $\dfrac{(6 \times 16)}{180} \times 100 = 53.3\%$

Practise these calculations with Questions 5 and 6 at the end of this unit. If you are not sure whether you have done them correctly, convert the formula you have obtained back into percentage composition and see if you get back to the same percentages!

Answers to diagnostic test

If you score less than 80%, then work through the text and re-test yourself at the end using this same test. If you still get a low score then try the unit at a later date.

1 The relative atomic mass of xenon is 131.3   **(1)**
2 58.5 (23 from sodium + 35.5 from chlorine)   **(1)**
3 10 g. The RFM of calcium carbonate is 100 and the RAM of Ca is 40. Therefore, calcumn constitutes $^{40}/_{100}$ths of the mass of calcium carbonate. That is 40 g in 100 g, and 10 g in 25 g, or $40/100 \times 25 = 10$ g.   **(4)**
4 10 g. The RFM of calcium carbonate is 100. So 1 mole would have a mass of 100 g and 0.1 mol has a mass of $0.1 \times 100 = 10$ g.   **(4)**
5 0.585 g. RFM for NaCl = 58.5 and RFM for $AgNO_3$ = 170. 1.7 g of $AgNO_3$ = 0.01 mol (from eqn. 4.1.1). This reaction requires 1 mol of NaCl for each mol of $AgNO_3$, so 0.01 mol of NaCl is required. Convert mol into mass by multiplying by the RFM. 0.01 mol = $0.01 \times 58.5 = 0.585$ g.   **(4)**
6 79.3%.
   Percentage yield = actual yield/theoretical yield $\times$ 100
                    = 6.9/8.7 $\times$ 100= 79.3%   **(2)**
7 CH_2O **(4)**

Element	C	H	O
Mass or %	40	6.7	53.3
RAM	12	1	16
No. of moles (mass/RAM)	3.33	6.7	3.33
Divide by smallest value (3.33)	1	2	1
Atom ratio	1	2	1

$\therefore$ Empirical formula is CH_2O

20 Marks   (80% = 16)

FURTHER QUESTIONS (ANSWERS GIVEN IN APPENDIX)

Questions to give you extra practice with the types of calculations shown in this unit. All RAMs can be found in the periodic table on page ii at the front of the book.

1 2.0 g of calcium react with 8.0 g of bromine to give 10 g of calcium bromide. How much calcium bromide will be produced if 6.6 g of calcium reacts with 26.4 g of bromine?
2 What is the mass of 1 mole of manganese(II) fluoride, MnF_2?
3 How many molecules are there in 234.8 g of silver iodide?
4 Magnesium bromide reacts with silver nitrate to give solid silver bromide and a solution of magnesium nitrate according to the reaction:

$MgBr_2(aq) + AgNO_3(aq) \rightarrow AgBr(s) + Mg(NO_3)_2(aq)$

a) Balance the equation.
b) 10 g of silver nitrate will yield 11.06 g of silver bromide. How

that the concentration of the solution is $5/58.5 \times 1/0.075$ mol dm^{-3}. Again, putting this into an equation gives us:

$$\text{concentration} = \frac{\text{mass in g}}{\text{RMM or RFM} \times \text{vol in dm}^{-3}} = \frac{5}{58.5} \times \frac{1}{0.075}$$

$$= 1.14 \text{ mol dm}^{-3}$$

d) A 700 cm^3 bottle of wine contains 70 g of ethanol. Ethanol, C_2H_5OH, has an RMM of 46. Let us put this one directly into an equation.

$$\text{No. of moles of ethanol} = \frac{\text{mass in g}}{\text{RMM}} = \frac{70}{46}$$

$$\text{Volume} = 0.7 \text{ dm}^3$$

$$\therefore \text{concentration} = \frac{\text{no. of moles}}{\text{vol. in dm}^3} = \frac{70}{46 \times 0.7} = 2.17 \text{ mol dm}^{-3}$$

Example 2:
Calculate the mass of the solute when the concentration is known.

In these examples, you are told the concentration of the solution and you have to calculate the amount of solute present in the solution.

a) In 1 dm^3 of a 2.5 mol dm^{-3} solution of sodium chloride (RFM 58.5), how much salt (in g) is in this saline solution? We start with the expression for concentration:

$$\text{concentration} = \frac{\text{mass in g}}{\text{RFM} \times \text{vol. in dm}^3 \text{ in which it is dissolved}} \quad (4.4.2)$$

We need to know the mass in g, so we rearrange the equation so that 'mass' is on its own on the LHS. This then becomes:

mass = concentration $\times$ RFM $\times$ volume

In this example, the concentration is 2.5 mol dm^{-3}, the volume of liquid is 1 dm^3, and the RFM of sodium chloride is 58.5.

mass = concentration $\times$ RFM $\times$ vol
mass = $2.5 \times 58.5 \times 1 = 146.25$ g

b) 100 cm^3 of a 2.5 mol dm^{-3} solution of sodium chloride (RFM 58.5). The mass in grammes is given by the concentration multiplied by the RFM times the volume in dm^3. The concentration is still 2.5 mol dm^{-3}, and the RFM still 58.5, but the volume is only 0.1 dm^3.

mass = concentration $\times$ RFM $\times$ vol
mass = $2.5 \times 58.5 \times 0.1 = 14.625$ g

c) In this next example, you are asked how much substance is needed to make up a solution of a particular concentration. Although this sounds like a different sort of question, it uses exactly the same equation as the last question.
   How many g of sugar (sucrose, $C_{12}H_{22}O_{11}$, RMM 342) are needed to make 250 cm^3 of a 0.01 mol dm^{-3} solution? We are given the

volume (0.250 dm³), the concentration (0.01 mol dm⁻³) and the relative molecular mass. The 'unknown' in the equation is the mass in grammes, and this is what we want on the LHS of the equation. Using the same equation we used in example (b):

mass = concentration × RMM × vol
mass = 0.01 × 342 × 0.25 = 0.855 g

Example 3:
Volume required, knowing the mass and concentration.

Sometimes we may have a 'stock' solution on the laboratory shelves. The solution is of a particular concentration, and we need to know what volume of solution to take to give a required amount of substance, either in grammes or moles. Calculate what volume of a solution is required to provide:

a) 2.5 g of sodium sulfate using a 2M (2 mol dm⁻³) solution. From a standard 2M solution of sodium sulfate, how many cm³ of solution do we need to use to provide 2.5 g?
   Sodium sulfate is Na_2SO_4. In this example we are given the mass in grammes (2.5), the concentration (2 mol dm⁻³), and the RFM (142). The 'unknown' is the volume in dm³.
   Rearrange the standard equation: concentration = mass in grammes divided by (RFM × volume), so that volume is on its own on the LHS.

$$\text{concentration} = \frac{\text{mass in g}}{\text{RFM} \times \text{vol in dm}^{-3}}$$

$$\text{volume} = \frac{\text{mass in g}}{\text{RFM} \times \text{concentration}}$$

$$\text{volume} = \frac{2.5}{142 \times 2} = 0.0088 \text{ dm}^{-3} = 8.8 \text{ cm}^3$$

b) We have a standard 4M (4 mol dm⁻³) solution of sodium hydroxide (RFM, 40). How much of this solution must we take to give 17 g of NaOH? We are given the concentration (4 mol dm⁻³), the mass in grammes (17 g), and the RFM (40). The unknown is again the volume.
   Rearrange the standard equation: concentration = mass in grammes divided by RFM × volume so that volume is on its own on the LHS.

$$\text{concentration} = \frac{\text{mass in g}}{\text{RFM} \times \text{vol in dm}^{-3}}$$

$$\text{volume} = \frac{\text{g}}{\text{RFM} \times \text{concentration}} = \frac{17}{40 \times 4} = 0.106 \text{ dm}^3 = 106 \text{ cm}^3$$

Further examples from this and all following sections can be found at the end of the unit.

4.4.2 Diluting Solutions

In our normal lives we are used to dealing with concentrated solutions and diluting them to achieve the required concentration. Bottles of squash come in a concentrated form that must be diluted before drinking,

Answers to diagnostic test

If you score less than 80%, then work through the unit and re-test yourself at the end using this same test. If you still get a low score then re-work the unit at a later date.

1 40 g. The RFM of NaOH is 40, so 1 mole has a mass of 40 g.
    1 dm³ of a 1 mol dm⁻³ solution contains 1 mole = 40 g.   **(4)**

2 Add 3 dm³ of water. This will bring the total volume to
    4 dm³, but the solution will still contain 2 moles of NaOH.
    The solution will now be 0.5 mol dm⁻³.   **(4)**

3 RMM of ethanoic acid is 60. No. of moles of ethanoic acid
    is 4/60 = 0.067. Concentration = moles/volume
    = 0.067/0.1 dm³ = 0.67 mol dm⁻³.   **(4)**

4 1 dm³ of a 6 mol dm⁻³ of sulfuric acid contains 6 moles of
    acid. One-sixth of 1 dm³ of a 6 mol dm⁻³ solution of sulfuric
    acid contains 1 mole of acid. So 1.75/6 dm³ of a 6 mol dm⁻³
    solution of sulfuric acid = 291.7 cm³ are required.   **(4)**

5 RFM of Na₂CO₃ = 106. 1 dm³ of a 1 mol dm⁻³ solution
    contains 106 g. So 100 cm³ contains 10.6 g.   **(4)**

20 Marks   (80% = 16)

FURTHER QUESTIONS (ANSWERS GIVEN IN APPENDIX)

1 Calculate the concentration in mol dm⁻³ of the following solutions:

   a) 9.5 g of sodium hydroxide (NaOH) in 450 cm³ of aqueous solution.

   b) 3.64 mg sodium chloride (NaCl) in 60 cm³ of aqueous solution.

   c) 20 g of barium chloride (BaCl₂) in 1.2 dm³ of solution.

2 Calculate what volume of the stated solution would be required to provide the following:

   a) 6 g of sodium chloride (NaCl) from a 0.1 mol dm⁻³ solution.

   b) 12.0 g of calcium chloride (CaCl₂) from a 6 mol dm⁻³ solution.

   c) 75 g of ethanol (C₂H₅OH) from a 2 mol dm⁻³ solution.

3 Calculate how many grammes of solute are necessary to produce the following solutions:

   a) 250 cm³ of a 0.025 mol dm⁻³ solution of calcium chloride (CaCl₂).

   b) 120 cm³ of a 0.15 mol dm⁻³ solution of glucose (C₆H₁₂O₆).

   c) 55 cm³ of a 1.125 mol dm⁻³ solution of sodium chloride (NaCl).

4 If concentrated sulfuric acid has a concentration of 17.8 mol dm⁻³, calculate the concentration of 500 cm³ of a solution of sulfuric acid prepared by diluting 25 cm³ of the concentrated acid.

Unit 4.5
Gases

Starter check

What you need to know before starting this unit.

You need to be familiar with the Kelvin temperature scale and the relationship between mass, volume and density. You should also have covered the relationship between mass and moles in Unit 4.3, and the explanation of pressure in Unit 1.1.

Aims **By the end of this unit you should:**

- Understand the general characteristics of gases.
- Be familiar with the units used for measuring the pressure of gases.
- Understand the relationship between the pressure, temperature and volume of a gas, and be able to solve problems using that relationship.
- Know the characteristics of an ideal gas and the rules that govern its behaviour (kinetic theory of gases).
- Understand vapour pressure.

Diagnostic test

Try this test at the start of the unit. If you score more than 80%, then use this unit as a revision for yourself and scan through the text. If you score less than 80%, then work through the text and re-test yourself at the end using this same test.

The answers are at the end of the unit.

1 Which of the following is not an observed property of a gas? **(1)**

a) Varies its shape and volume to fit the container.
b) Expands indefinitely and fills the space uniformly.
c) Compresses infinitely and uniformly.
d) Forms a homogeneous mixture with other gases.
e) Has a low density.

2 A gas has a volume of 275×10^{-3} dm^3 at 1.0 atm and 288 K. What will be its volume at 10.0 atm and 373 K? [Hint, always give answers in the units specified in the question. 1 atm $= 1 \times 10^5$ N m$^{-2} = 1 \times 10^5$ Pa (Pascal)]. **(3)**

3 How many grammes of nitrogen are in a 50 dm^3 cylinder at 288 K and 1.5 atm pressure? [RAM: N, 14; 1 mole of gas occupies 24 dm^3 at 298 K and 1 atm (1×10^5 Pa)]. **(3)**

4 Which two variables concerning gases are connected by Boyle's Law? **(1)**

5 A mixture of gases has the following partial pressures for the component gases at 373 K in a volume of 5 dm³: oxygen, 2.08×10^4 Pa; nitrogen, 8.05×10^4 Pa. What is the total pressure of the mixture? **(2)**

10 Marks (80% = 8)

4.5.1 Introduction

Elements and compounds exist in solid, liquid or gaseous states.

A substance in the gaseous state has neither fixed shape nor volume and spreads itself evenly throughout the whole of the container. If a gas is released into a room, it distributes itself eventually throughout the room – we mostly detect gases by our sense of smell – think of the aroma of freshly made bread that fills a bakery, or the rich scent of freshly brewed coffee. Every gas possesses the ability to move through any available space, even if that space is occupied by another gas, and even if there seems to be no wind or movement to spread the gas. This property is called **diffusion**, and is explained by assuming that gases consist of very small particles that are capable of moving very fast. When we use the term 'gas', we normally refer to any substance that is in the gaseous state at normal temperatures and pressures. We can, of course, convert almost any solid or liquid to a gas simply by changing the conditions of temperature and pressure.

Watching a hot air balloon take shape on the ground and finally rise up into the air lifting a large basket and several people is a fascinating sight (Figure 4.5.1).

Figure 4.5.1
Hot air balloon

It seems impossible that a light fabric bag can lift such a weight. It does so because the air inside the fabric bag or balloon is hotter than the outside air and the density of the heated air is much lower. Children's helium balloons 'float' attached to their strings because the density of helium at room temperature and pressure is less than that of air. Release them from their strings and they will float off high into the sky – causing great distress to the children!

The gaseous state is one of the three common states of matter; it is the simplest state, and is the one that scientists know most about.

Over the 300 years that scientists have been studying gases systematically, they have found that all gases have the following five general characteristics:

1 A gas has no definite shape, but takes the shape of its the container.
2 A gas will expand uniformly to occupy all the space in that container.
3 A gas can be compressed so that a large amount can be contained in a small container, *e.g.* a diver's pressurised tanks may contain enough air for several hours' breathing. The gas cannot be compressed indefinitely

though; as the particles of gas get closer together the gas becomes harder to compress and, below a particular temperature which varies from gas to gas, pressure converts a gas to a liquid (see Unit 1.1).

4 The densities of gases are much lower than those of solids or liquids.

5 When two or more gases are together in the same container, they will mix completely and uniformly (we say that the gas mixture is homogeneous).

Before we can understand the laws that govern the way gases behave, we need to know the units and ways in which gases are measured.

4.5.2 Units

Although SI units are standardised for pressure and volume, different units are used in different circumstances. Read a question carefully to see which units are used and use the same ones for the answer unless a different unit is specified.

PRESSURE

We have said that a gas occupies the volume of its container. If we put more gas into the same container, the volume stays the same. What changes is the pressure. Gas pressure results from the molecules or atoms of gas being in constant motion and hitting the walls of their container. The more often collisions occur between the gas particles and the wall, the higher the pressure. Fewer collisions means lower pressure. All gases exert some pressure, even if it is very small. See also Section 1.1.7 on page 11 for a further discussion of pressure.

The gases in the air exert a pressure on their 'container', the Earth's surface. Atmospheric pressure is the pressure exerted by air molecules colliding with their environment. Atmospheric pressure can be measured using a mercury barometer, the earliest form of which was devised in 1643 by Torricelli. It consists of a long glass tube with one end closed, filled with mercury and held upside down with its open end in a dish of mercury open to the air (see Figure 4.5.2). Torricelli found that whatever diameter the glass tube had, the level of mercury in the tube always fell to roughly the same level above the mercury in the dish, assuming that the experiment was carried out at the same height above sea level and at the same temperature. At sea level the height of the column of mercury was around 76 cm. The pressure of gas which at sea level and 0 °C (*i.e.* 273 K), supports a column of mercury 76.0 cm high is called **standard atmospheric pressure**.

Standard pressure may be expressed in other units given in Table 4.5.1.

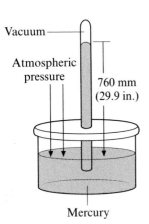

Figure 4.5.2
Torricellian barometer

Vacuum

Atmospheric pressure

760 mm
(29.9 in.)

Mercury

Table 4.5.1 *Table of units for pressure measurement*

Unit	Standard pressure
Centimetres of mercury	76.0 cm
Millimetres of mercury	760 mm
Inches of mercury	29.9 ins
Torr	760 torr
Pounds per square inch (psi)	14.7 psi
Atmospheres	1 atm
millibars	1013 mbr
Pascals	1.013×10^5 Pa* (1 Pa = 1 N m^{-2})

*Often simplified to 1×10^5 Pa.

The Pascal (or $N\,m^{-2}$) is the standard unit of pressure and should be used in chemistry, but the others are still frequently used in weather forecasting, for measuring tyre pressures, or on barometers.

Atmospheric pressure is always quoted:
- at sea level (at higher altitudes, the air is thinner and pressure is lower);
- at 0 °C or 273 K (at higher temperatures, the particles of gas are moving faster and the pressure will be greater).

VOLUME

The volume of a gas is measured in m^3, dm^3 or cm^3, where $1\ m^3 = 1000\ dm^3$; $1\ dm^3 = 1000\ cm^3$. Another common unit is the litre (l), where $1\ l = 10^3\ cm^3 = 1\ dm^3$.

TEMPERATURE

Gas calculations are always worked out in Kelvin, abbreviated to K. (You do not need to use the degree (°) sign with Kelvin, just the letter K.) The Kelvin temperature scale is produced by adding 273 to the temperature in degrees C. The lowest temperature on the Kelvin scale, 0 K, represents *absolute zero* or −273 °C. A few useful conversions are given in Table 4.5.2.

Table 4.5.2 *Conversion table for Kelvin temperature scale*

	°C	K
Absolute zero	−273	0
Freezing point of water	0	273
Standard room temperature	25	298
Boiling point of water	100	373

STANDARD TEMPERATURE AND PRESSURE (STP)

This is 25 °C or 298 K and 1 atm or 1.013×10^5 Pa pressure.

Standard measurements of gases are always given at STP. If no information is given, it will be assumed that the conditions are STP.

MOLAR VOLUME

We have already seen (Section 4.3.1) that a mole of any substance, whether the particles are atoms, molecules or formula units, contains 6.02×10^{23} 'particles'. What does this mean when we are talking about gases? How can we judge 1 mole of gas? Experiments have shown that at STP, *i.e.* at 25 °C and 1 atmosphere pressure,

1 mole of gas will occupy a volume of 24 dm^3.
At 0 °C and 1 atm, 1 mole of gas occupies 22.4 dm^3.

By using this fact in combination with the relationship *density = mass/volume*, we can work out how much gas in moles or in grammes is present in any volume under different conditions. It is probably easiest to show this using an example.

Example:
A helium balloon has a volume of 5 dm^3. How many moles of helium are in the balloon?

At STP, 1 mole of helium occupies a volume of 24 dm³. The balloon has a volume of 5 dm³, so the number of moles of helium = 5/24 moles = 0.208 moles.

4.5.3 Kinetic Theory

The word **kinetic** comes from the Greek word which means 'to do with movement'.

In kinetic theory, energy in the forms of heat and movement are related; higher temperature indicates greater movement. All particles are assumed to move, whether they are in solids, liquids or gases.

- Solids. The amount of movement in solids is very small and only occurs as vibration within the lattice of particles in the solid.
- Liquids. The particles of liquid move at random. (They flow, but are attracted together in a mass.)
- Gases. The particles in gases have movement that is limited only by their container. When kinetic theory is applied to gases, the following assumptions are made:
  1 In a container of gas, the space occupied by the gas molecules is very small compared with the total space. So when we think of the volume of a gas, we are mostly considering empty space. This explains why gases have low density and can be compressed.
  2 The molecules of gas move about continuously and rapidly, hitting each other and the walls of the container randomly. This explains why different gases mix completely. Collisions between the gas molecules and the walls account for the pressure exerted by the gas.
  3 These collisions cause no damage to the gas molecules, which continue to move and collide again and again. These collisions are known as elastic collisions. The result of elastic collisions is that, although energy may be transferred from one molecule to another during the collision, the system as a whole loses no energy as a result of the movement and collisions.
  4 The individual gas molecules have different kinetic energies as a result of these collisions, but the average kinetic energy of the gas molecules (*i.e.* the total energy of the system divided by number of molecules) is proportional to the temperature in Kelvin, and is the same at a set temperature for all gases. So, at the same temperature and pressure, equal volumes of nitrogen and oxygen or any other gas will have the same amount of kinetic energy. This means that 'heavier' gas molecules (*e.g.* CO_2, SO_2) will, at a given temperature, on average be moving more slowly than 'light' gas molecules (*e.g.* H_2, He). At absolute zero, 0 K, all vibration and movement ceases, and substances are considered to have zero energy.
  5 The molecules of gas have no forces of attraction between them. The movement of the molecules is random and independent of the other molecules present. (As we shall see, this is only an approximation. If there were no attractions at all, no gas could *ever* be liquefied.)

4.5.4 Ideal Gases and the Gas Laws

The concept of an ideal gas is one that perfectly fulfils the above assumptions and has the following characteristics:

1 The volume of the gas is negligible compared to the volume of the container.

2 Collisions are perfectly elastic.

3 There are no attractive or repulsive forces between the gas molecules.

Real gases such as nitrogen and oxygen are not truly ideal, but as long as we avoid very extreme conditions of temperature and pressure, we can assume that real gases will behave as ideal gases and will follow the gas laws.

There are three main factors that affect gases: **temperature**, **pressure** and **volume**.

We will look at three possible cases in which we keep one of these factors constant and see what effect changing a second factor has on the third.

BOYLE'S LAW: THE EFFECT OF PRESSURE ON THE VOLUME OF A GAS AT CONSTANT TEMPERATURE

Robert Boyle's experiments showed that gas volume went down as the pressure in the container went up; or in scientific terms, *the volume of a set mass of gas at constant temperature is inversely proportional to the pressure*. Or, using symbols:

$$V \propto \frac{1}{P} \text{ or, introducing a constant, } V = k \times \frac{1}{P}$$

So, if the pressure is doubled, the volume will be halved. As the volume is decreased, collisions will occur more often and the pressure will increase. It is not easy to work with equations with proportional signs, and we have no way (here!) of knowing what the value of the constant is. So, in order to make use of this equation, we can measure the temperature and pressure at two different temperature values and put them into the equation:

$$\text{At temperature 1: } V_1 = k \times \frac{1}{P_1}$$

$$\text{At temperature 2: } V_2 = k \times \frac{1}{P_2}$$

We now rearrange both the equations (to rearrange an equation, see Unit 4.1) to put the constant on its own on one side of the equals sign:

$$P_1 V_1 = k \text{ and } P_2 V_2 = k$$

but if k is a constant it is always the same and so can be cancelled from both sides. This gives equation 4.5.1:

$$P_1 V_1 = P_2 V_2 \tag{4.5.1}$$

CHARLES'S LAW: THE EFFECT OF TEMPERATURE ON THE VOLUME OF A GAS AT CONSTANT PRESSURE

Jacques Charles had a very practical interest in the effect of temperature on volume – he was an early hot air balloonist! Gay-Lussac was a French scientist who performed experiments with a practical outcome in mind. He too was a hot air balloonist who held the world balloon altitude record for many years. He also came to a similar conclusion about the relationship between the temperature and volume of a fixed mass of gas at a constant pressure.

They found that the volume of a gas varied with the temperature. Once the Kelvin scale was established in the 19th century, it could be said that *the volume of a fixed mass of gas at constant pressure is directly proportional to the Kelvin temperature.*

$V \propto T$ or, introducing a constant, $V = k \times T$

So if a gas is heated by 1 °C, its volume increased by 1/273 of its volume at 0 °C. (The 273 comes from the conversion of Celsius to Kelvin.) The opposite is also true: if the temperature is decreased by 1 °C, then the volume decreases by 1/273. In theory, at absolute zero (-273 °C) the volume of a gas is 0 dm^3, but real gases turn into liquids before this point.

To make this into a useful equation, we need to rearrange the equation as before, and measure under two sets of conditions to get rid of the unknown constant. If we do this we get equation 4.5.2:

$$\frac{V_1}{T_1} = \frac{V_2}{T_2} \qquad\qquad (4.5.2)$$

THE EFFECT OF TEMPERATURE ON THE PRESSURE OF A GAS AT CONSTANT VOLUME

A 1 °C rise in temperature results in a 1/273 rise in pressure. In other words, *pressure for a fixed mass of gas at constant volume is directly proportional to temperature.*

$P \propto T$ or, introducing a constant, $P = k \times T$

Rearranging the equation as before, we get equation 4.5.3:

$$\frac{P_1}{T_1} = \frac{P_2}{T_2} \qquad\qquad (4.5.3)$$

When we started looking at these laws, we made the assumption that we could keep one of the factors constant and vary the other two. In practice, however, this is not what happens. All three are likely to vary together.

THE COMBINED GAS LAW

We can combine all three of the above expressions (equations 4.5.1–4.5.3) in the form of a combined gas law (equation 4.5.4):

$$\frac{P_1 V_1}{T_1} = \frac{P_2 V_2}{T_2} \qquad\qquad (4.5.4)$$

Let us see how this combined gas law can be used in a practical situation.

Example 1:
A sample of gas has a volume of 5 dm^3 at atmospheric pressure and a temperature of 25 °C. What is its volume at 0 °C? (Remember, atmospheric pressure and 25 °C are the standard temperature and pressure, or STP.)

Answer: In this case, the 'unknown' of the equation is the second volume, so we will rearrange the equation to put that on the left-hand side; the pressure is constant throughout, so $P_1 = P_2$. So:

$$\frac{P_1 V_1}{T_1} = \frac{P_2 V_2}{T_2}$$

$$\frac{P_1 V_1 \times T_2}{T_1 \times P_1} = V_2$$

$P_1 = P_2$, so they cancel out, leaving $\dfrac{V_1 \times T_2}{T_1} = V_2$

Don't forget that temperature is in Kelvin!
(*Always* convert temperatures in gas calculations to Kelvin. In this case, 25 °C = 298 K.)

$$\frac{5.0 \times 273}{298} = V_2$$

$$V_2 = 4.58 \text{ dm}^3$$

Example 2:
A gas has a volume of 250 cm³ at STP. What will its pressure become if the volume is changed to 600 cm³? In this question, the temperature is constant throughout, and the second pressure is the unknown. The two volumes are 0.25 and 0.6 dm³ and the first pressure is 1 atm. It may be easier to see this if we arrange the figures as in the table below:

P_1	P_2	V_1	V_2	$T_1 = T_2$
1 atm (= 1 × 10⁵ Pa)	unknown	0.25 dm³	0.60 dm³	25 °C = 298 K

In this question the units are given in atmospheres, so we will answer it in the same units. The answer in standard units is given for completeness. Let's rearrange the combined gas equation to give P_2 on the LHS and put these figures in.

$$\frac{P_1 V_1}{T_1} = \frac{P_2 V_2}{T_2}$$

$$\frac{P_1 V_1 \times T_2}{T_1 \times V_2} = P_2$$

$$P_2 = \frac{1 \times 2.5 \times 10^{-4} \times 298}{298 \times 6.0 \times 10^{-4}} = 0.42 \text{ atm } (= 4.2 \times 10^4 \text{ Pa})$$

Example 3:
Suppose we have the same situation as in the last question, but we also allow the temperature to rise to 50 °C. What will the new pressure be then? Let us use a table again:

P_1	P_2	V_1	V_2	T_1	T_2
1 atm (1 × 10⁵ Pa)	unknown	0.25 dm³	0.60 dm³	25 °C = 298 K	50 °C = 323 K

Put these figures in the rearranged gas equation. Alternative SI units are again given for comparison.

$$\frac{P_1 V_1}{T_1} = \frac{P_2 V_2}{T_2}$$

$$\frac{P_1 V_1 \times T_2}{T_1 \times V_2} = P_2$$

$$P_2 = \frac{1 \times 0.25 \times 323}{298 \times 0.60} = 0.45 \text{ atm } (= 4.5 \times 10^4 \text{ Pa})$$

4.5.5 The Ideal Gas Equation – Calculating the Mass and Relative Molecular Mass of a Gas

The ideal gas equation combines the variables of temperature, pressure and volume that we have been dealing with in the previous sections, but also allows us to calculate the mass in either grammes or moles and also an approximate molar mass for the particular gas. The previous gas laws involved an unknown constant that we eliminated from the calculation by taking temperatures, *etc.*, at two different levels. In the ideal gas equation, we are introduced to the *universal gas constant*, R, which enables us to do the measurements under one set of conditions only. The difficulty arising from this is that the units of the gas constant are dependent on the units in which the other variables are measured, so it is important to think about the units you are working in. A selection of values for R using different units is listed in Table 4.5.3.

Table 4.5.3 *Values for the universal gas constant*

Units of pressure	Units of volume	Value of R	Units of R
Pascals (Pa)	m^3	8.314	$J\, K^{-1}\, mol^{-1}$*
$N\, m^{-2}$	m^3	8.314	$N\, m\, K^{-1}\, mol^{-1}$
Atmospheres (atm)	dm^3	0.082	$atm\, dm^3\, K^{-1}\, mol^{-1}$
Torr	dm^3	62.3	$torr\, dm^3\, K^{-1}\, mol^{-1}$

*These are the units of R used in modern texts – but as usual, take your lead from the units in the question.

Hints on working with the ideal gas equation:

- Temperatures *must* be in **K**, but the questions will probably have temperatures in °C. Make sure you add '273' to the values in °C.
- Volumes will probably be in dm^3 or cm^3 in the questions. Make sure you correlate the volume given in the question with the unit of volume used in the value of R you are using (see Table 4.5.3). To convert dm^3 to m^3, simply multiply by '10^{-3}', and cm^3 to dm^3, multiply by 10^{-3} or for cm^3 to m^3, by 10^{-6}. Use the handy conversion chart shown below:

1 cm³ = 10⁻³ dm³		= 10⁻⁶ m³
1 dm³ = 1000 cm³ (10³ cm³)		= 10⁻³ m³
1 m³ = 10⁶ cm³		= 1000 dm³ (10³ dm³)

$$\times 10^{-3} \qquad \times 10^{-3} \qquad \times 10^{3} \qquad \times 10^{3}$$
$$cm^3 \;\rightarrow\; dm^3 \;\rightarrow\; m^3 \;\rightarrow\; dm^3 \;\rightarrow\; cm^3$$

- The universal gas constant has units that depend on the units of pressure. Use the table to make sure you use the correct value.
- Read the question carefully! Assume the units needed for the answer are those used in the question unless it says otherwise. In the example given below two different systems of units are given – you only need to use the set given in a question.

The ideal gas equation is expressed in a mathematical form as equation 4.5.5:

$$PV = nRT \tag{4.5.5}$$

	SI units
P = Pressure	Pa
V = Volume	m³
T = Temperature	K
n = amount of gas	mol
R = universal gas constant	J K⁻¹ mol⁻¹

Using this equation, we can calculate the number of moles of gas in a given container.

Example:
A 10 dm³ (10×10^{-3} m³ = 10^{-2} m³) cylinder contains oxygen at 27 °C (= 300 K) and 5 atm (= 5×10^5 Pa) pressure. How many moles of oxygen does this contain?

Start with the universal gas equation, $PV = nRT$ and rearrange it so that the unknown is on the LHS.

$$n = \frac{PV}{RT}$$

As the question is set, the pressure is in atmospheres, so we need to use the appropriate value of the universal gas constant, $R = 0.082$ atm dm³ K⁻¹ mol⁻¹. Substitute the figures into the equation:

$$n = \frac{PV}{RT} = \frac{5 \times 10}{0.082 \times 300} = 2.03 \text{ moles of oxygen}$$

(If we repeat this calculation using SI units, the volume and temperature have the same units but the pressure is 5.065×10^5 Pa, so we need to use the value of R that relates to Pascals; but we get exactly the same answer, as the units cancel out!)

In SI units,

$$n = \frac{PV}{RT} = \frac{5 \times 10^5 \times 10 \times 10^{-3}}{8.314 \times 300} = 2 \text{ moles of oxygen}$$

This can then be converted to mass in grammes using equation 4.1.1: mass = number of moles × RMM (see Unit 4.3). In this case, mass in grammes = $2.03 \times$ RMM of O_2 = $2.03 \times 32 = 64.96$ g.

To calculate the relative molecular mass of a gas, we need to measure

- the volume of gas;
- the mass of gas in that volume;
- the temperature of the gas.

The ideal gas equation then enables the calculation of:

- the number of moles of gas;
- the relative molecular mass (from equation 4.1.1, RMM = mass/ moles).

4.5.6 Changing the Density of a Gas

Next time you watch a science fiction or horror film, watch for the film maker creating an atmosphere of horror by having his characters walking through a swirling mass of mist. In fact he is quite literally creating an 'atmosphere', usually of water droplets. Solid carbon dioxide, obtained in frozen blocks, evaporates directly to a gas without passing through a liquid phase (it *sublimes*) and the density of the gaseous carbon dioxide is greater than that of air. The gas initially sinks to the ground, and as its temperature is below 0 °C, it condenses any water vapour in the air surrounding it and gives the impression of mist. Notice that it is the water droplets that you actually see as mist, not the carbon dioxide which is, of course, a colourless, odourless gas. Since the volume of a gas varies with temperature and pressure, so will its density.

4.5.7 Partial Pressures

In all the previous examples we have looked at, we have only considered pure gases. But what happens if we have a mixture of gases? The kinetic theory of gases says that most of the space inside a container of gas is empty – the gas molecules themselves occupy very little of it. Also, gases are said by the theory to have no forces of attraction between them. A mixture of gases can therefore occupy the same volume as one gas on its own. However, because there are more molecules of gas present, there will be more collisions between the gases and the walls of the container, so the pressure will be higher. The total pressure arising from the separate pressures caused by each gas in a mixture of gases is summarised by Dalton's Law of Partial pressures.

This law states that each gas in a mixture of gases exerts a pressure that is the same as if it were the only gas in the container. This is called a **partial pressure**.

The total pressure of the mixture, *i.e.* the total pressure in the container, is equal to the sum of the partial pressures of all the gases present.

Putting this in the form of an example; if a vessel holds three gases, A, B, and C, and the partial pressure of each is P_a, P_b, and P_c, respectively, then the total pressure in the vessel is:

$$P = P_a + P_b + P_c \qquad (4.5.6)$$

Answers to diagnostic test

If you score less than 80%, then work through the text and re-test yourself at the end using the same test. If you still get a low score then re-work the unit at a later date.

1 The answer is (c) – gases cannot be compressed indefinitely.   **(1)**

2 3.56×10^{-3} dm³. Use the combined gas law and insert the figures; $V_1 = 275 \times 10^{-3}$ atm, $P_1 = 1.0$ and $T_1 = 288$ K. V_2 is the unknown, $P_2 = 10.0$ atm and $T_2 = 373$ K.

$$\frac{P_1 V_1}{T_1} = \frac{P_2 V_2}{T_2}$$

$$\frac{1.0 \times 275 \times 10^{-3}}{288} = \frac{10.0 \times V_2}{373}$$

$$\frac{1.0 \times 275 \times 10^{-3} \times 373}{288 \times 10.0} = V_2$$

$$V_2 = 35.6 \times 10^{-3} \text{ dm}^3 = 3.56 \times 10^{-2} \text{ dm}^3 \qquad \textbf{(3)}$$

3 90.44 g. First use the above combined gas law to see what volume the gas would occupy at 298 K and 1 atm. In this case, $V_1 = 50$ dm³, $T_1 = 288$ K, and $P_1 = 1.5$ atm, $T_2 = 298$ K and $P_2 = 1.0$ atm.

$$\frac{P_1 V_1}{T_1} = \frac{P_2 V_2}{T_2}$$

$$\frac{1.5 \times 50}{288} = \frac{1.0 \times V_2}{298}$$

$$\frac{1.50 \times 50 \times 298}{288 \times 1.0} = V_2$$

$$V_2 = 77.60 \text{ dm}^3$$

If 1 mole of gas occupies 24 dm³, then 77.6 dm³ must contain $\frac{77.6}{24}$ moles $= 3.23$ mol

No. of moles $= \dfrac{\text{mass in grammes,}}{\text{RMM}}$

so mass in grammes $=$ no. of moles $\times$ RMM $= 3.23 \times 28$

$\therefore$ mass of nitrogen $= 90.44$ g   **(3)**

4 Boyle's Law gives a relationship for the change of volume with pressure at constant temperature.   **(1)**

5 1.013×10^5 Pa. Add together the partial pressures of the two gases, $2.08 \times 10^4 + 8.05 \times 10^4 = 10.13 \times 10^4 = 1.013 \times 10^5$ Pa.   **(2)**

10 Marks   (80% = 8)

Unit 4.6
General Properties of Solutions

Starter check

What you need to know before starting this unit.

You need to be familiar with the Kelvin temperature scale and the relationship between mass, volume and density. You should also have covered the relationship between mass and moles in Unit 4.3.

Aims **By the end of this unit you should:**

- Know the factors that influence solubility and rate of solution.
- Be familiar with different types of solution.
- Understand colligative properties.
- Be familiar with the concept of colloids.
- Be familiar with the principles of recrystallisation.

Diagnostic test

Try this test at the start of the unit. If you score more than 80%, then use this unit as a revision for yourself and scan through the text. If you score less than 80%, then work through the text and re-test yourself at the end using this same test.

The answers are at the end of the unit.

1 A solution of salt in water is heated to boiling point. At what temperature would you expect the solution to boil?

   a) At 100 °C;
   b) below 100 °C;
   c) above 100 °C. **(2)**

2 Would you expect glycerine [propane-1,2,3-triol, $C_3H_5(OH)_3$] to be miscible with water? **(2)**

3 Would you expect potassium nitrate (KNO_3) to be more soluble in hot water than in cold? **(2)**

4 Barium sulfate is very soluble in water. True or false? **(2)**

5 Colloids contain particles that reflect light. True or false? **(2)**

10 Marks (80% = 8)

4.6.1 Introduction

Solutions are something we all take for granted in our everyday lives. As you read this, you may be preparing a solution – by stirring the coffee into which you have just put a spoonful of sugar! Even the coffee itself is partly a solution – of roasted coffee bean extract in hot water. If you think about some aspects of your life, you will find that you know quite a lot about solutions and their properties, it's just that you don't normally think of them in terms of chemistry. What about that cup of instant coffee? Have you ever poured water in from the kettle only to find that the kettle hadn't been switched on and the water was still cold? What happened? Probably the coffee granules floated on the top of the water in a disgusting manner. So coffee is more soluble in hot than in cold water. What about that bottle of milk that was left out of the fridge when you went away? Apart from the rancid smell, the protein and fat separate from the water (which is the main component of milk) and form a layer on the top. The components are no longer mixed together. What happened last winter when a cold snap of weather caught you out before you had put antifreeze in the car – so the car wouldn't start? Antifreeze lowers the temperature at which water freezes, and so stops the water that cools the engine freezing. Putting salt on the roads also lowers the freezing point of water; this helps to keep the road free of ice. In this unit we will look at solutions and think about their properties generally.

What is a solution? A solution is defined as an homogeneous mixture involving two or more components. When a solid dissolves in a liquid, the solid seems to 'disappear'. The visibly-sized particles of solid break up into tiny particles that are distributed evenly throughout the liquid, and the liquid remains transparent. The solute forms a sort of bond with the solvent. In the case of aqueous solutions, this bonding could be hydrogen bonding, as with sugar, or 'hydration', as with sodium chloride (see below for an explanation of hydration). A solution consists of:

NEVER confuse 'dissolving' with 'melting'. What is the essential difference?

- *A solvent.* This is the component that is present in the greatest quantity. It is frequently, but not necessarily, a liquid. If a solution consists of a liquid and a solid or a liquid and a gas, then, conventionally, the liquid will be the solvent even if it is not the major component of the mixture.
- *A solute.* This is the substance dissolved in the solution. It is usually in smaller amounts than the volume of solvent.

Other terms used about solutions are:

- *Insoluble.* When a solid is added to a liquid (solvent), if the solid is insoluble it will sink to the bottom of the container and remain essentially unchanged physically. No 'bond' is formed between the liquid and the solid and it will not dissolve. (N.B. Water will dissolve *tiny* amounts of even the most apparently insoluble material. Every time you drink a glass of water, you drink a tiny amount of the glass.)
- *Suspension.* Sometimes when an insoluble solid is added to a liquid and the mixture is shaken, the particles of solid don't sink but remain distributed throughout the solvent; and they do this without changing in physical appearance. The solution goes 'cloudy' – that is, it is no longer transparent because of the solid particles distributed in it. Milk is a suspension of fat droplets in water. Mist is a suspension of droplets of water in air. Smoke is a suspension of tiny solid particles in air.
- *Slightly soluble.* Few solutes are completely insoluble in a solvent. Often a small amount dissolves. When this amount is measurable, the solute is said to be partially or slightly soluble.

- *Miscible.* The above terms refer principally to solutions of solids in liquids. When one liquid dissolves in another liquid to form a solution, the term miscible is used. The solvent will be the component that is present in the larger quantity. Ethanol is completely miscible with water – a property used to produce the wide strength range of alcoholic drinks.
- *Immiscible.* When one liquid is insoluble in another liquid, *i.e.* the liquids do not mix, the term used is immiscible. The two liquids will form separate layers with the one with the lower density on top. For example, a puddle in the street often has an iridescent or rainbow-like appearance due to a thin layer of oil on the top. Oil is immiscible with water, and has the lower density, so it floats (see Figure 4.6.1).
- *Partially miscible.* If one liquid dissolves in a second liquid to a certain extent, then the two layers will be formed again. This time the layers will consist of:
  a) a solution of the more dense liquid in the less dense liquid. The less dense liquid forms the upper layer;
  b) a solution of the less dense liquid in the more dense liquid. The more dense liquid forms the lower layer.
  An example of this is a mixture of ethane-1,2-diol (glycol, ethylene glycol, antifreeze) and trichoromethane (chloroform) shown in Figure 4.6.2.

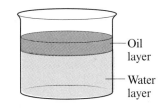
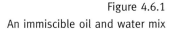

Figure 4.6.1
An immiscible oil and water mix

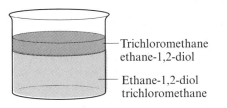

Figure 4.6.2
A partially miscible mixture

- *Solvation and hydration*
  In the introduction to this unit, we talked about a 'bond' being formed between the solute and solvent. What is this bond?

  In Unit 2.3 we learned about hydrogen bonds. These are inter-molecular electrostatic forces of attraction between certain polar molecules (often water). When an ionic solid is stirred into water, the polar water molecules surround the particles and electrostatic bonds are formed between the oxygen and the metal ion and also between the hydrogen and the anion. These bonds help the solid to dissolve and break into individual ions (Figure 4.6.3). This process is called **hydration**.

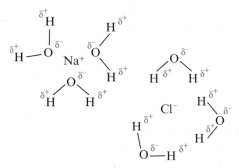

Figure 4.6.3
Hydration of sodium chloride

Some hydrogen bonds between the water molecules will be broken in this process, before the new bonds between the water and the ions can be formed. Because breaking and forming bonds involves energy absorption and release, there is likely to be a temperature change noticeable in the temperature of the solution. Some salts, when added to water, result in the solution warming up; whereas most salts result in a reduction in temperature. Whether the solution warms up (heat given out, exothermic) or cools down (heat taken in, endothermic) – see Unit 5.1 for a discussion of bond energy – depends on whether the heat required to break the hydrogen bonds is less than (for an exothermic change) or greater than (for an endothermic one) the heat given out by the new bonds forming.

Although hydration is a very common process; the solvent doesn't have to be water. It can occur when any polar solute is dissolved in a polar solvent. The general term for this is **solvation**.

Hydration is thus a special form of solvation.

4.6.2 Types of Solution

Although we think of solutions as solids (or solutes) dissolved in liquids as solvents, other phases can form solutions. Table 4.6.1 shows examples of the different types.

Table 4.6.1 *Types of solution*

Solvent	Solute	Example
Liquid	Gas	Fizzy drinks (*e.g.* carbon dioxide in water)
Liquid	Liquid	Alcoholic drinks (ethanol in water)
		Antifreeze (ethylene glycol in water)
Liquid	Solid	Salt or sugar in water
Solid	Solid	Metal alloys such as brass (zinc in copper), 9 ct gold (gold in copper)

4.6.3 Factors Affecting Solubility

When two components are mixed together, what controls whether they will form a solution, and how fast the solution will be formed? The solubility of a solute in a solvent is governed by:

PROPERTIES OF THE SOLUTE AND SOLVENT

This takes us back to the different types of bonding in the two components (Units 2.2 and 2.3). A compound containing ionic bonds will usually dissolve in a solvent which is also polar (see Section 2.3.4) and not in a solvent with purely covalent bonding. Similarly, a compound which is covalent will usually dissolve in a covalently-bonded organic solvent. Thus sodium chloride, an ionic compound, will dissolve in water, which is polar, but not in petrol, which contains no polar bonds. Sodium chloride will have slight solubility in some organic solvents that have a small amount of ionic character, such as ethanol, due to solvation. We can summarise solubility in the phrase:

Like dissolves like!

TEMPERATURE

For the great majority of solutions of solids in liquids and liquids in liquids, increasing the temperature usually increases the solubility. Indeed, lead chloride is insoluble in cold water, but soluble in hot water.

But for solutions of gases in water, however, increasing the temperature has the opposite effect. A bottle of a fizzy drink will lose its sparkling quality if kept at room temperature; it keeps longer in the fridge. The solubility of air in water also decreases as the temperature rises; this can have serious consequences for fish, which find less dissolved oxygen in the ponds during the heat of the day than at night. Fish therefore tend to move to deeper water during the day, as the shallow warmer water at the edge is more denuded of oxygen.

PRESSURE

Pressure effects are only of importance in the case of gases in liquids. Increasing the pressure of the gas leads to increased solubility. Fizzy drinks are under quite a lot of pressure; the container has to withstand the desire of the dissolved gas to escape, particularly during transport, when the bottles get shaken about and possibly subjected to extremes of heat and cold. Champagne is stored in thick glass bottles with special corks. These restrict the volume and also prevent the gas leaking out over prolonged storage times. Cola is stored in plastic bottles, but will not keep fizzy for years in the way that champagne does.

Formula 1 race winners shake the bottle before loosening the cork . . .

4.6.4 Factors Affecting Rate of Solution

TEMPERATURE

Because substances at higher temperature have more kinetic energy, they will go into solution faster. Particles with more energy will undergo more collisions and move about faster and further through the solvent. Thus the solution will become homogeneous faster.

RATE OF STIRRING

Increases the speed of dissolving. The solid particles would otherwise soon be surrounded by saturated solution, which would stop more solid dissolving until fresh solvent particles diffused in. The energy input from the hand holding the spoon transfers to moving the particles through the solution. A stirred cup of coffee dissolves the sugar faster!

PARTICLE SIZE

Solutions consist of individual particles, molecules and ions dispersed through a solvent. The 'lumps' of solute therefore have to be split up into individual particles. The act of dissolving takes place at the surface of the 'lump' and smaller 'lumps' have a larger surface area:volume ratio. The smaller the 'lumps' are to start with, the faster they will dissolve. So powdered salt will dissolve faster than salt with coarse grains. The effect of particle size on the rate of a chemical reaction is discussed in Section 5.3.4.

4.6.5 Ionic Salts – Solubility Rules

Although it is possible to deduce the solubility of common inorganic compounds, this following list of solubilities is useful for quick reference.

Rules for the solubility of inorganic substances in water:

1 All nitrates (NO_3^-) and most ethanoates, (acetates, CH_3COO^-) are soluble.
2 All common chlorides are soluble except for silver chloride (AgCl); mercury(II) chloride ($HgCl_2$); and lead(II) chloride ($PbCl_2$), although the latter is soluble in hot water.
3 All common sulfates are soluble, except for barium sulfate ($BaSO_4$) and lead sulfate ($PbSO_4$). [Calcium sulfate ($CaSO_4$) and silver sulfate (Ag_2SO_4) are only slightly soluble.]
4 All common salts of the metals of Group 1 in the periodic table – sodium, potassium (Na, K) – and all common ammonium salts (NH_4^+) are soluble.
5 All common acids are soluble.
6 All oxides and hydroxides are **in**soluble except those of Group 1 (Li, Na, K). The hydroxides of Group 2 in the periodic table become a little more soluble as we move down the group, calcium $\rightarrow$ strontium $\rightarrow$ barium (Ca, Sr, Ba).
7 All sulfides are insoluble except those of Group 1 (Li, Na, K) and ammonium.
8 All carbonates are insoluble except for sodium, potassium and ammonium.

N.B. Some salts, *e.g.* aluminium chloride, decompose – sometimes quite violently – with water. But do not bother with these at this stage!

4.6.6 Recrystallisation

Suppose you have carried out a chemical reaction and the product and the starting material are both solids. If the reaction has gone to completion and you have no starting material remaining, then it may be quite pure. But the probability is that the product will contain some impurities, perhaps even a lot of starting material. How do you separate the compound you want from the others present? Recrystallisation is a routine technique for doing this. It is used for purifying a crystalline solid which contains fairly small amounts of impurities.

First choose your solvent. The choice of solvent is crucial. You need a solvent in which your chosen compound is only slightly soluble when cold, but very soluble when the solvent is hot. In addition, the impurities present must be either completely insoluble in the solvent, both cold or hot, or stay dissolved in the solvent even when it is cold. The solvents that are required for purifying specific mixtures are often mixtures of two or three components in set proportions.

Steps for recrystallisation:

1 You will need a supply of hot solvent and a hot flask (usually a conical flask) with a hot funnel and a folded filter paper.
2 Dissolve the mixture in the *minimum* of boiling solvent.
3 Some impurities may not be soluble in the solvent. If this is the case,

pour the solution and solid impurities through the hot funnel and filter paper. The impurities will stay behind in the filter paper. (The funnel must be hot, or your compound will crystallise out in the funnel stem – a disaster!)

4 If the solution is coloured because of other impurities, it can be cleaned up using a small amount of decolourising charcoal. Add this to the solution, shake, and again filter through a *hot* funnel to remove the charcoal.

5 Allow the solution to cool. Your compound should crystallise out of the liquid, leaving any soluble impurities behind. If traces of impurities remain, recrystallise at least once more.

6 Maximum yield can often be achieved by sitting the flask in an ice bath.

4.6.7 Colligative Properties

There are some properties of solutions which are common to all solutions with non-volatile solutes, and which vary only with the concentration of particles added to the solution and do not depend on the identity of the particles. These properties are called **colligative properties**. The most important of these properties are:

- The decrease of vapour pressure above the solvent.
- The raising of the boiling point of the solvent.
- The lowering of the freezing point of the solvent.
- The osmotic pressure of the solution.

These properties can only be accurately predicted for dilute solutions.

VAPOUR PRESSURE

In a pure liquid, there is a tendency for some of the molecules of solvent to escape from the liquid and pass into the vapour phase. That is, the molecules form an atmosphere of gas above the liquid. Eventually equilibrium will be reached, with molecules evaporating from the surface at a rate equal to the rate at which they are condensing back into the liquid. This equilibrium pressure of the vapour above the liquid (partial pressure, see Section 4.5.7) is called the **vapour pressure**. Each solvent will have its own vapour pressure, which will increase with temperature. If the liquid is a mixture, both components contribute to the vapour pressure depending on their relative concentrations, and the total vapour pressure will be the sum of the two.
  So let us imagine a mass of pure water.

- The individual vapour pressure of water is W.
  The total vapour pressure above the mass of water will be $W \times 100\%$.
- If the mixture has a molecule ratio of 90% water and 10% ethanol, (ethanol has a vapour pressure of, say, E) then:
  The total vapour pressure of the mixture will be $(90\% \times W + 10\% \times E)$.
- If the liquid contains dissolved solute that is not volatile (like salt or sugar), the solute will not contribute to the vapour pressure.
  So, the total vapour pressure will only be due to the liquid.
- OK, let us go back to the mass of water, with its total vapour pressure of $100\% \times W$.

Let us make the mixture (in terms of ratios of particles) 90% water (vapour pressure, $90\% \times W$) and 10% salt.

- The vapour pressure of salt is zero because it is not volatile.
  So, the total vapour pressure will be (90% × W + 10% × 0) = 90% × W.
- Dissolving an involatile solute in a solution will thus reduce the vapour pressure.

RAISING THE BOILING POINT

The vapour pressure of a liquid increases with increasing temperature. When the vapour pressure of the liquid reaches the external pressure on the surface of the liquid, usually atmospheric pressure, the liquid boils. When a non-volatile solute is dissolved in the liquid, the vapour pressure is lowered and particles of solvent find it harder to escape from the liquid surface. The solution has to be heated more to push the vapour pressure up to the outside pressure, leading to a boiling point for the solution which is higher than that of the pure solvent.

LOWERING THE FREEZING POINT

During the freezing of a solvent, the solvent particles lose kinetic energy and the attraction between the particles becomes relatively more important. Eventually the solvent turns into a solid, with the properties of a solid. When a non-volatile solute is added, particles of this get in the way, stopping solvent particles sticking together. Thus the solvent has to lose more energy; that is, the temperature has to be lowered further to allow freezing to take place.

(This description is a simplification, but it gives the general picture!)

OSMOTIC PRESSURE

Consider two solutions separated by a membrane that will allow solvent particles through, but *not* solute particles. (This is called a semi-permeable membrane, and they are very important in biology – and in you! Think about why your kidneys are there, for a start.) All particles are moving, so solvent particles can hit the membrane and pass through it. Pretty obviously, the solution that has the greater concentration of *solvent* particles – *i.e.* the less concentrated solution (the dilute one with fewer solute particles in it) will have more solvent particles going through the membrane.

Solvent particles go through the membrane in both directions, but they go through faster from the more dilute solution into the more concentrated one. That is **osmosis** – the overall passage of solvent from a more dilute solution into a more concentrated one (see Figure 4.6.4). It is a natural consequence of the fact that particles in liquids are in constant movement.

The **osmotic pressure** of a solution is the pressure that has to be exerted on the more concentrated *solution* side of a semi-permeable membrane to *just* stop overall passage of solvent into it from *pure solvent* on the other side of the membrane.

The solution with the higher concentration of solute is said to be 'hypertonic', while the solution with the lower concentration is 'hypotonic'. You have probably come across these terms, hyper- and hypo-, in other areas, such as the slang term for being over-excitable – a person is described as being hyper! They actually originate from old Greek words now used in medical terms; hyper- means excessive and hypo- means under. So 'hypertension' is high blood pressure and 'hypotension' is low

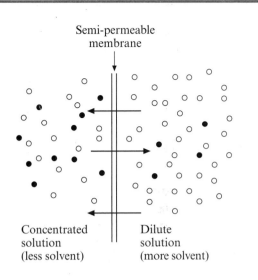

Figure 4.6.4
Osmosis

blood pressure. When the concentrations of the two solutions are equalised by overall passage of solvent through the membrane from the more dilute solution, the solutions are said to be 'isotonic'. (Iso- means 'the same'.)

One type of semi-permeable membrane is usually made of cellophane these days, but originally it was made of pig's intestines! It allows solvent to pass through it and *small* particles of solute such as salt, but not large molecules (so salt, glucose, amino acids or water would pass through the intestine wall, but not molecules of protein or starch). Similarly, water, urea and ionic materials can be removed from blood by the kidneys.

There are other types of semi-permeable membrane that will not allow charged particles such as Na^+ or K^+ to pass through, but will allow through uncharged particles of almost any size. A lot of biology involves selective membranes of one kind or another.

Figure 4.6.5 shows a semi-permeable membrane in a laboratory version of dialysis, a sort of imitation of kidney function. We have a concentrated solution of sugar, say, inside the membrane and the dilute solution, or pure solvent, outside. The two solutions are in contact with each other, and over a period of time, the volume of liquid inside the semi-permeable membrane increases as, overall, solvent moves inside it. It is unlikely that the concentrations will be completely equalised.

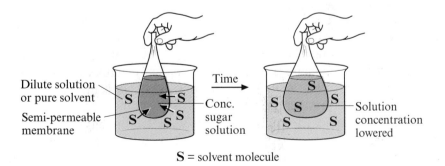

S = solvent molecule

Figure 4.6.5
Laboratory experiment to demonstrate dialysis

Like all the other colligative properties, the osmotic pressure depends on the amount of solute present in the solution, not its identity.

4.6.8 Colloidal Dispersions

Colloids, or colloidal dispersions, are special sorts of suspension. Because they are not true solutions we do not use the words solute and solvent when talking about colloids, but 'dispersed phase' and 'dispersion medium'. The individual dispersed particles are not bound to the dispersing medium in the way that solutes are in solutions, and they do not settle out of the solution, in the way that suspensions eventually do. Colloids have dispersed particle sizes that are bigger than individual molecules, typically 10^{-9} to 10^{-6} m in diameter. You may be able to see the particles in a microscope, particularly by their scattering of light; and, unlike a true solution, the colloid has a cloudy or milky appearance. This is often described as 'turbid'.

A colloidal dispersion can sometimes be made by violently shaking the solute and solvent together. Electrostatic forces of repulsion caused by ions of the same charge adsorbed on their surface keep the dispersed particles in small associated groups. They also prevent the dispersed particles from coming together or 'aggregating' and settling or precipitating out of the dispersing medium.

The dispersed particles can be removed from the solution by centrifuging, a technique that involves spinning a tube round at high speed, when the heavier particles are forced to the bottom of the tube. Another way of removing the dispersed particles is to add an ionic solute. Triply-charged aluminium (Al^{3+}) ions are particularly effective at coagulating colloids. Aluminium sulfate is used both in sewage treatment, and in after-shave (to stop bleeding from nicks . . .). The sensitivity of colloidal dispersions to ions is also why fine mud occurs in estuaries – the finest particles are carried in the river water and are deposited when they come into contact with the ions in sea water.

When a beam of light is passed through a colloid, the particles scatter the light and appear in the beam as bright specks of light. This occurs because the dispersed particles are large enough to 'reflect' or scatter the light, producing a visible beam. The smaller particles of solute in a solution are too small to cause visible reflection. This property of colloids is called the Tyndall effect. You have probably observed the Tyndall effect – when the sun breaks through the clouds, the rays of sunlight that you see are shown up because they are scattered by dust particles in the air.

When the colloid is looked at under a microscope, the reflections of light from the individual dispersed particles show that the particles move in a random fashion through the liquid, due to being continuously bombarded by particles of the dispersion medium. This constant bombardment also helps prevents the dispersed particles from settling out. A botanist called Robert Brown, who was trying to study pollen grains suspended in water, first observed this phenomenon. When he put the suspension under the microscope, he found that instead of being able to see his grains more clearly, they wouldn't keep still but danced about all over the place! This random movement was called Brownian motion, after him.

Like solutions, dispersed systems do not have to be solids in liquids. Table 4.6.2 lists some different types of colloids that we commonly encounter, probably without knowing that they are colloids. Blood is a colloidal dispersion!

Table 4.6.2 *Types of colloid*

Dispersion Medium	Dispersed Phase	Name	Example
Gas	Liquid	Liquid aerosol	Mist, clouds
Gas	Slid	Solid aerosol	Smoke
Liquid	Gas	Foam	Fire-fighting foam
Liquid	Immiscible liquid	Emulsion	Milk, mayonnaise
Liquid	Solid	Sol	Latex paint or glue
Solid	Gas	Solid foam	Polyurethane or expanded polystyrene foams
Solid	Liquid	Gel	'Instant Whip', hair gel, jelly
Solid	Solid	Solid sol	Coloured glass, opals

Unit 5.1
Energy Changes

Starter check

What you need to know before you start this unit.

You will need to know about the structure of the atom (Module 1) and ionic and covalent bonding (Module 2).

Aims **By the end of this unit you should understand and be able to:**

- Appreciate the ideas of reaction kinetics and the factors that affect rates of reactions.
- Begin to appreciate the theories about why reaction rates differ.
- Appreciate the ideas of energy changes in reactions, and use the terms 'endothermic' and 'exothermic'.

Diagnostic test

Try this test at the start of the unit. If you score more than 80%, then use this unit as a revision for yourself and scan through the text. If you score less than 80%, then work through the text and re-test yourself at the end by using this same test.

The answers are at the end of the unit.

1 Which reactions, in solution, are quicker, ionic or covalent? **(1)**
2 In each of the following test tube reactions how could you tell if a chemical reaction had occurred?

   a) Silver nitrate solution and hydrochloric acid are added together.
   b) Magnesium metal is added to hydrochloric acid.
   c) Water is added to quicklime (calcium oxide). **(3)**
3 The rate of a chemical reaction increases when the temperature of the system is raised. Why is this? **(2)**
4 What is meant by the 'activation energy' of a reaction? **(2)**
5 What is meant by the terms 'exothermic' and 'endothermic'? **(2)**
6 Some reactions are affected by catalysts.

   a) Explain what a catalyst is and give an example. **(2)**
   b) Describe briefly how catalysts are believed to work. **(2)**

7 a) What does the symbol ΔH represent?

   b) What do you understand by 'ΔH is negative'.

   c) Would ΔH be positive or negative for the reaction in
      Question 2(c)?   **(3)**

8 a) More energy is released when strong bonds are formed than
      when weaker bonds are made. TRUE or FALSE?   **(1)**

   b) So long as particles collide, they will react together. TRUE or
      FALSE? Explain your answer.   **(2)**

20 Marks   (80% = 16)

5.1.1 Introduction to Energy Changes

You are a walking, talking chemical reaction (Figure 5.1.1). Every step you take, every thought and breath, involves a series of physiological changes that makes use of chemical reactions – and you don't even have to think about it! The energy needed to ensure that these reactions occur comes from the body creating it from the food that we eat (Figure 5.1.2).

Figure 5.1.1
Many chemical reactions occur in the body

Figure 5.1.2
Food provides energy for the chemical reactions

Fats and carbohydrates are important biological fuels – the chemical energy stored in these foods can be converted into heat and movement energy, electrical energy in brain and muscles, and of course sound energy. Protein foods are essential for tissue growth and cell repair, although they will also provide energy for the body if eaten in excess of basic nutritional requirements.

These are just some of the reactions that are going on in your body, but there are many more situations in which chemical reactions occur. The chemist in industry plays a vital role ensuring that processes are economically viable, both in terms of production yields and time taken. Every reaction is different; it is necessary to be able to distinguish between them, to understand the optimum conditions for effective outcomes, and to be able to measure how quickly each one proceeds. The following units in this module have been written to develop your background knowledge and understanding of these basic ideas.

5.1.2 Energy Changes in Chemical Reactions

- When a piece of magnesium metal is burned in air it produces a lot of heat and a brilliant white light; in older cameras this was seen when the 'flashgun' was used. The same reaction is now used to provide bright light in emergency 'flares' and fireworks.

$$2Mg(s) + O_2(g) \rightarrow 2MgO(s)$$

- When a mixture of hydrogen and oxygen gases is ignited the noise heard varies from a 'squeaky pop' to a deafening bang depending on the quantities present.

$$2H_2(g) + O_2(g) \rightarrow 2H_2O(l)$$

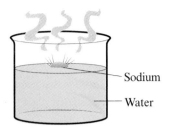

Figure 5.1.3
The exothermic reaction of sodium metal and water

• If a small piece of sodium is placed in a bowl of water (Figure 5.1.3), it whizzes around the surface and melts into a silvery globule, it may even explode and burn with a yellow flame.

$$2Na(s) + 2H_2O(l) \rightarrow 2NaOH(aq) + H_2(g)$$

In these examples, some of the chemical energy that was stored within the bond structure of the reactants is given out during the reaction as heat or light, and even as movement or sound. All reactions involve energy changes and often energy is released, as heat or light for example, and more rarely some reactions absorb energy during their course. This unit considers various chemical reactions in which there is a noticeable or measurable amount of energy released or absorbed; it also provides a theoretical basis on which to build a picture as to why reactions have such different outcomes.

Have you ever thought about:

• *Why magnesium burns so fiercely when lit? (After all, some alloys used in building aircraft contain magnesium and they do not burst into flames, unless they are heated* very *strongly.)*
• *The significance of the flame or spark in the reaction between hydrogen and oxygen?*
• *Why sodium reacts so spectacularly with water whereas iron rusts quite slowly?*

5.1.3 Collision Theory

All matter is made up of particles, and in the study of **reaction kinetics** we are interested in the behaviour of atoms, molecules and ions and how they react together. It seems very obvious to say this, but two chemicals can only react if their particles come into contact with each other.

The particles in gases and liquids are in constant motion and consequently they will collide with each other. Millions of such collisions occur every second, but not every one produces a new substance; if they did reactions would be over in less than a millionth of a second. We have much everyday evidence to show that this is not so; for instance, if iron rusted on its first contact with the air then bridges and towers could never be built, so evidently not all collisions are effective.

Imagine a playground full of young school children. They are more likely to bump into each other when they are running around than when they are walking or sitting. Sometimes they will run past each other, sometimes they may just brush sleeves and stumble slightly, but occasionally two children will collide head-on! The difference in outcomes is both obvious and painful: the more children in the playground, the greater the likelihood of a collision. Likewise, in a chemical reaction the type of collision that occurs between particles, and how 'hard' or energetic it is, will help decide whether or not a reaction will happen. Also, just as in the overcrowded playground, the, more particles there are in a given volume – the greater their **concentration** – the greater the chance of collision.

We can use a simple reaction to illustrate this. Hydrogen and oxygen are both diatomic gases. Under the right conditions they can be made to react to form water. Figure 5.1.4 illustrates the way in which different molecules might collide.

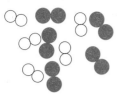

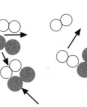

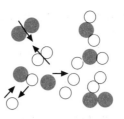

The molecules move about randomly and at different speeds. Some may collide.

Some molecules only catch 'glancing blows' – low energy collisions with too little energy to break bonds. So no reaction!

'Head-on' collisions, real 'knock-outs' – break bonds in the hydrogen and oxygen molecules, allowing a new product to be formed.

Figure 5.1.4
Different types of collision in a mixture of gases

When two particles collide, 'head-on', with sufficient speed, the collision is a 'high energy' one and the reaction occurs quickly. Imagine two sports cars in a head-on collision! For instance, when the temperature is increased for a reacting system, the particles, on average, move faster – they have a greater kinetic energy (movement energy) and collisions are 'harder'. So, **increasing the temperature increases the rate at which reactions occur**.

SUMMARY

In order for a reaction to occur particles must collide with each other – the more particles that are in a certain amount of space, that is a fixed volume of gas or solution, and the faster they are moving, then the greater the likelihood of collisions and so a reaction.

Increases in both temperature and concentration increase the rate of reaction: temperature has the greater effect – to see why this should be, read on!

5.1.4 Making and Breaking Bonds

All reactions involve the breaking of bonds, between the atoms or ions in the reactants, and the formation of new bonds to form different substances, *i.e.* the products. Atoms, ions and molecules involved in chemical reactions have two types of energy associated them:

- **kinetic energy** – because they are moving;
- **bond energy** – because there are forces between the atoms or ions; these forces are the bonds that hold the atoms or ions together (see Units 2.1 and 2.2).

Energy is needed to break particles apart – to break bonds or overcome attractions. This energy is usually supplied when particles collide. However, if this is insufficient then more energy must be transferred somehow, and this is usually in the form of *heat*.

Strong bonds will need more energy to break them than weak bonds.

Coal and wood are both well known as domestic fuels. They do not burn without help, however – they both must be heated to fairly high temperatures before they will 'catch fire'. Only then can the chemicals in the fuels react with the air. Once started, the reaction continues, giving out much more heat, which in turn makes more fuel hot enough to ignite. The original input of heat has to be supplied in order to break apart the bonds between atoms of oxygen molecules, O_2, in the air, and the carbon and hydrogen atoms, *etc.*, that make up the chemicals in wood or coal. This initial supply of heat energy has been **taken in**, *i.e.* **absorbed by the reactants**. However, we are well aware that both coal and wood produce large amounts of heat when burned. This *excess heat* is given out as new bonds are formed in the making of carbon dioxide and water *etc.* as the products of combustion.

The **energy diagram** (Figure 5.1.5) represents what is happening during the reaction as bonds are broken when methane gas (*e.g.* as ordinary North Sea gas used for cooking) is burned in air. It can be seen that the energy content of the system increases – **bond breaking needs energy**.

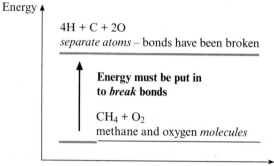

Figure 5.1.5
Energy diagram for the breaking of bonds in the oxidation of methane

$$CH_4(g) + 2O_2(g) \rightarrow CO_2(g) + 2H_2O(g)$$

The next stage of the reaction involves the making of new bonds as carbon dioxide and water are formed. It has already been said that, 'the stronger the bond the greater the energy needed to break it'; conversely, when bonds are formed, energy is *released*:

The stronger the bond formed the more energy is released – making bonds gives out energy.

The energy diagram for the formation of new bonds is shown in Figure 5.1.6. It can be seen that the energy content of the system falls as the bonds are made.

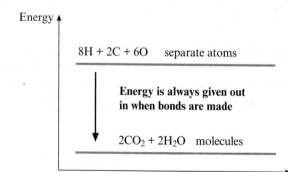

Figure 5.1.6
Energy diagram for the formation of new bonds in the oxidation of methane

5.1.5 Exothermic and Endothermic Reactions

Chemical changes are always accompanied by *changes* in the energy content of the materials that are reacting, and the *change* is usually observed in the form of heat. Indeed, in many cases the change in temperature when substances react is the only evidence that a chemical reaction has taken place. Changes in the heat content of a system – called the **heat of reaction** – can be measured by the *change in the temperature* of the system during the course of the reaction. The symbol for 'heat of reaction' is ΔH (where Δ is the Greek letter *delta*; used in mathematics to mean 'difference').[1]

EXOTHERMIC REACTIONS

a) When magnesium is burned in oxygen it burns with a white hot flame.

$$2Mg(s) + O_2(g) \rightarrow 2MgO(s) \qquad \Delta H \text{ is negative}$$

The stored chemical energy in the magnesium and oxygen is partly transferred to stored chemical energy in the bonds of magnesium oxide while the rest is evolved as heat and light. Thus, the magnesium oxide has a lower overall energy content than the elements from which it is made.

b) When magnesium ribbon is added to dilute hydrochloric acid, you can feel the test tube getting hot! HEAT is given out.

$$Mg(s) + 2HCl(aq) \rightarrow MgCl_2(aq) + H_2(g) \qquad \Delta H \text{ is negative}$$

c) When water is added to white anhydrous copper sulfate powder, to form the hydrated blue solid, the reaction gives out heat and there is a rise in temperature until enough water has been added and the reaction stops.

$$CuSO_4(s) + 5H_2O(l) \rightarrow CuSO_4 \cdot 5H_2O(s) \qquad \Delta H \text{ is negative}$$
white   blue

All the reactions above give out energy to their surroundings, in particular, HEAT ENERGY.

- Reactions that **give out heat energy** are called **exothermic** reactions.
- By *convention* ΔH for an exothermic reaction is given a **negative** sign, showing that **heat** is **given out** – *i.e.* heat energy has been lost from the reacting system to the surroundings – the system ends up with a *lower* energy content.

---

[1] It is more accurate to say that ΔH represents the enthalpy change of a reaction – see Section 5.1.6.

The general energy diagram for an exothermic reaction is shown in Figure 5.1.7.

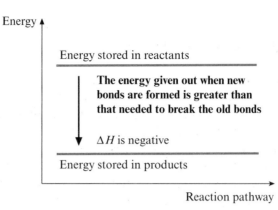

Figure 5.1.7
Energy diagram for an
exothermic reaction

The reactions above all demonstrate *energy changes*.

The Law of Conservation of Energy states that 'energy can neither be created nor destroyed – it only *changes* from one form to another'.

This process can also be shown as a word equation for an exothermic reaction.

Energy stored in the bonds of of the reactants = **Energy stored in the bonds of products + HEAT energy given OUT**

A huge number and variety of chemical reactions are exothermic and have energy diagrams like that in Figure 5.1.7 – some examples are given in Figure 5.1.8.

- *Respiration*
  $C_6H_{12}O_6(aq) + 6O_2(g) \rightarrow 6CO_2(g) + 6H_2O(l)$ ΔH is negative

- *Neutralisation*: acid + base → salt + water
  e.g. $H_2SO_4(aq) + 2NaOH(aq) \rightarrow Na_2SO_4(aq) + 2H_2O(l)$
  ΔH is negative

- *Hydration*
  e.g. $CuSO_4(aq) + 5H_2O(l) \rightarrow CuSO_4{\cdot}5H_2O(s)$ ΔH is negative
  white powder   blue crystals

- *Combination reactions*
  e.g. $NH_3(g) + HCl(g) \rightarrow NH_4Cl(s)$ ΔH is negative

  $Fe(s) + S(s) \rightarrow FeS(s)$ ΔH is negative

Figure 5.1.8
Some common exothermic
reactions

ENDOTHERMIC CHANGES

Look at what happens in the following processes:

- The decomposition of calcium carbonate to calcium oxide and carbon dioxide needs a lot of heat energy.
- Some physical processes such as evaporation, dissolving and melting take in heat energy – they are endothermic processes.

- When ammonium nitrate is dissolved in water you can feel the test tube get **colder** as the dissolving process takes in heat from your hand and the surroundings.

So where does this absorbed heat energy go?

When salt is dissolved in water heat is absorbed from the surroundings. This heat energy is needed to overcome the forces of attraction between the sodium and chloride ions in the crystal lattice, so that they dissolve and are then free to move about in the solution.

In each of the examples above, *the energy content of the products is greater than that of the reactants*. It is worth noting, however, that *genuine*, spontaneous, endothermic, chemical reactions are relatively *rare*, and some examples are given in Figure 5.1.9.

- *Photosynthesis*
  $6CO_2(g) + 6H_2O(l) \rightarrow C_6H_{12}O_6(aq) + 6O_2(g)$ ΔH is positive

- *Dehydration*
  e.g. $CuSO_4 \cdot 5H_2O(s) \rightarrow CuSO_4(s) + 5H_2O(l)$ ΔH is positive

- *Decomposition*
  e.g. $NH_4Cl(s) \rightarrow NH_3(g) + HCl(g)$ ΔH is positive

  $ZnCO_3(s) \rightarrow ZnO(s) + CO_2(g)$ ΔH is positive

Figure 5.1.9
Some common endothermic reactions

When ENDOTHERMIC changes and reactions happen, heat is taken in from the surroundings.

By convention ΔH is given a **positive** sign for **endothermic** processes and reactions – *i.e.* heat energy is gained by the system from its surroundings and the system ends up with a higher energy content (Figure 5.1.10).

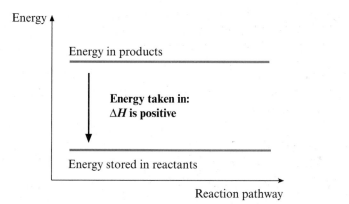

Figure 5.1.10
Energy diagram for an endothermic reaction

For an endothermic process this can be summarised by the word equation:

Energy stored in the bonds of **=** **Energy stored in the bonds of the**
of the products   **reactants + HEAT energy taken IN**

5.1.6 The Heat of Reaction

As stated above, the amount of energy given out or absorbed during a chemical reaction is called the **heat of reaction**: ΔH.

The heat of a reaction = **Energy content of the products − Energy content of the reactants**

In order to compare the amounts of energy gained or lost during a chemical reaction, it is necessary to have an agreed standard. The energy changes are measured as changes in the heat content of the reacting system, and so it is important that no energy is 'lost' in any other form. Experiments to measure heats of reaction are carried out in sealed containers surrounded by a known volume of water, so that all energy changes, even sound, are ultimately converted to heat changes which are measurable. The quantities of reactants used are based on those worked out from the balanced equation for the reaction. The heats of reaction have been experimentally determined for thousands of different reactions, and in each case the standard energy change is based on that involving **one mole** of reactant or product.

One mole of a chemical substance contains the same number of particles* as there are atoms in 12 g of the isotope carbon-12 (^{12}C). That number is 6.02×10^{23} particles (Avogadro's Constant).

For instance, the reaction between nitrogen and hydrogen to produce ammonia can be represented by the equation:

$$N_2(g) + 3H_2(g) \rightleftharpoons 2NH_3(g) \quad \Delta H = -92 \text{ kJ mol}^{-1}$$

But what is the heat of reaction for? One mole of ammonia or hydrogen or nitrogen?

The stoichiometric equation is the 'balanced' equation, *i.e.* that which shows the numerical relationship between the relative quantities (in terms of molecular mass) of substances in a reaction.

In order to be consistent we must state the heat of reaction with reference to the whole stoichiometric equation – so the standard heat of reaction for the production of one mole of ammonia is $\Delta H = -46 \text{ kJ mol}^{-1}$, based on:

$$\tfrac{1}{2}N_2(g) + \tfrac{3}{2}H_2(g) \rightleftharpoons NH_3(g) \quad \Delta H = -46 \text{ kJ mol}^{-1}$$

where mol^{-1} means '*per molar quantities indicated by the equation*'.

The stoichiometric equation has been '*re-balanced*' to show the molar quantities of hydrogen and nitrogen needed to produce *one mole of the product ammonia.*

You can read more about the use of the mole concept in Unit 4.3:

Taking the earlier example of burning methane gas in oxygen:

$$CH_4(g) + 2O_2(g) \rightarrow CO_2(g) + 2H_2O(l) \quad \Delta H = -890 \text{ kJ mol}^{-1}$$

When methane burns in air, *one mole* of methane combines with *two moles* of oxygen and 890 kilojoules (kJ) of heat is evolved.

---

* The particles might be atoms, molecules, ions, or formula units depending on the nature of the substance.

This is a *combustion reaction* and the heat of reaction, ΔH, is called the **heat of combustion**, ΔH_c, and is the amount of energy given out when one mole of methane burns in oxygen. In this case we would refer to **the heat of combustion of methane**, *i.e.* the heat change when one mole of methane is burned.

In practice, the data for heats of reaction are stated with reference to standard conditions of temperature and pressure, *i.e.* 298 K and 101.325 kPa (25 °C and 1 atm). If heats of reaction are measured under different conditions of temperature and pressure, it is likely that they will differ from the data.

5.1.7 Activation Energy

But why do some reactions occur 'spontaneously' whereas others need an energy boost?

We have seen that, for a chemical reaction to happen, molecules need not only to collide but must do so with sufficient energy. Breaking bonds needs energy; if molecules are moving too slowly and the collisions are weak, then there may not be sufficient energy to break the bonds and the reaction cannot proceed. The course of a chemical reaction might be compared to a pole-vault competition (Figure 5.1.11): only some competitors will have enough energy to get over the bar – others will need extra assistance or even the lowering of the bar, and some may never make it! The pole vaulter may improve performance by the use of technology to design a better pole or by drinking 'high energy' products. The equivalent may be achieved in a reaction by increasing the kinetic energy of the particles – making them move faster so that more of them have sufficient energy to get over the barrier. This can be done by heating the reactants (see Section 5.1.3).

Figure 5.1.11
Pole-vault

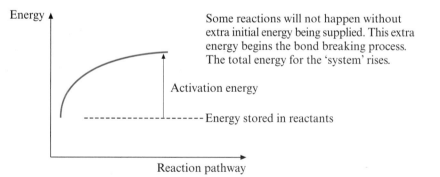

Some reactions will not happen without extra initial energy being supplied. This extra energy begins the bond breaking process. The total energy for the 'system' rises.

Figure 5.1.12
The energy barrier for a reaction – the activation energy

The minimum kinetic energy that particles must have before a reaction can take place is called the **activation energy** (Figures 5.1.12 and 5.1.13).

The activation energy, E_a, is the minimum kinetic energy needed to enable bonds to stretch and break so that atoms, ions or molecules can rearrange as the reaction proceeds and so form products.

More simply, it is the energy needed to break the *old* bonds before the *new* bonds can form.

Reactions that occur readily at room temperature have a relatively low activation energy, whereas those that need large inputs of heat have high activation energies. Increasing the temperature of a reacting system raises

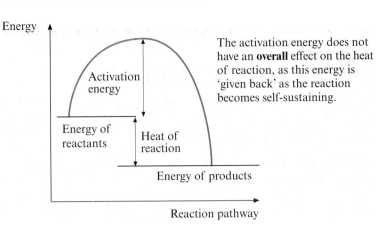

The activation energy does not have an **overall** effect on the heat of reaction, as this energy is 'given back' as the reaction becomes self-sustaining.

Figure 5.1.13
Energy diagram with summary of terms for a reaction

the kinetic energy of the particles so there are more high impact collisions which then result in a reaction.

Some reactions can be 'speeded up' by the use of catalysts. It is thought that the catalyst provides an alternative reaction pathway (mechanism) which has a lower activation energy (see Unit 5.3).

This could also be shown as in Figure 5.1.14:

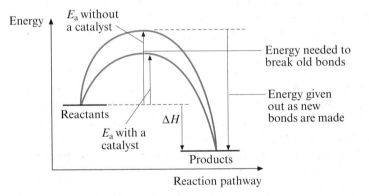

Figure 5.1.14
Speeding up a reaction by use of a catalyst

It can be seen that when an exothermic reaction occurs, the new bonds are stronger than the old ones. When an endothermic reaction occurs, the new bonds are weaker than the old ones.

Answers to diagnostic test

If you score less than 80%, then work through the text and re-test yourself at the end by using this same test. If you still have a low score then re-work the topic at a later date.

1 Ionic reactions are quicker than covalent ones.   **(1)**
2 a) You would see a precipitate (of silver chloride) form.
   b) You would see an effervescence (bubbling) as hydrogen
      was evolved *and/or* would feel the test tube get hot.
   c) You would feel the test tube get hotter.   **(3)**
3 When the temperature is increased the particles move faster
   and this makes them more likely to collide, and the collisions
   happen with greater energy. The increase in energy has by far
   the greater effect.   **(2)**
4 Activation energy is the minimum amount of kinetic energy
   that the particles must have in order to be able to react.   **(2)**
5 An exothermic reaction is one in which heat is given out by
   the reacting system to the surroundings – it gets hotter.
   Endothermic reactions need heat, which they draw from the
   surroundings – the reacting system gets colder.   **(2)**
6 a) A catalyst is a substance that speeds up the rate of a
      chemical reaction and can be recovered chemically
      unchanged at the end of the reaction. There are many,
      but manganese dioxide, platinum, and enzymes all
      catalyse different reactions.   **(2)**
   b) Catalysts are thought to work by allowing a reaction to
      proceed *via* a different route or pathway which has a
      lower activation energy, so more particles have sufficient
      kinetic energy to react.   **(2)**
7 a) ΔH is the symbol for the 'heat of reaction' or 'enthalpy
      of reaction'.
   b) When ΔH is negative the reaction gives out heat; it is
      exothermic.
   c) ΔH is negative.   **(3)**
8 a) True   **(1)**
   b) False – they must also have sufficient kinetic energy to
      overcome the energy barrier – the activation energy.   **(2)**

20 Marks   (80% = 16)

3 Consider the reaction between calcium carbonate (marble chips) and dilute hydrochloric acid.

$$CaCO_3(s) + 2HCl(aq) \rightarrow CaCl_2(aq) + CO_2(aq) + H_2O(l)$$

a) Sketch the graph showing how the volume of carbon dioxide gas evolved changes with time. .   **(3)**

b) On the same graph draw the line that shows what you would expect if the acid was twice the concentration as in (a).   **(2)**

4 In the reaction,

$$Mg(s) + H_2SO_4(aq) \rightarrow MgSO_4(aq) + H_2(g)$$

given that 48 cm³ of hydrogen gas was given off in 30 s:

a) What is the average rate of formation of hydrogen, in cm³ per second (cm³ s⁻¹)?   **(1)**

b) What is the average rate of formation of hydrogen in moles per second (mol s⁻¹)?   **(5)**

(1 mol of any gas occupies 24 dm³ at standard temperature and pressure)

25 Marks   (80% = 20)

5.2.1 Introduction

It is important to know how quickly a reaction occurs and how to change its rate, especially on an industrial scale, as slow processes are usually uneconomical and ways have to be found to speed them up.

REACTIONS HAPPEN AT DIFFERENT RATES

Chemical reactions take place at a variety of rates, for instance explosions are very rapid whereas iron rusts relatively slowly. Many ionic reactions in solutions happen in a fraction of a second and so they appear to be 'instantaneous'.

- When a solution of silver nitrate is added to solution of sodium chloride a white precipitate forms immediately. The precipitate is *silver chloride*.

  Sodium chloride + silver nitrate $\rightarrow$ sodium nitrate + silver chloride
  $Na^+Cl^-(aq) + Ag^+[NO_3]^-(aq) \rightarrow Na^+[NO_3]^-(aq) + Ag^+Cl^-(s)$

- Silver bromide, on light-sensitive photographic paper, reacts almost *instantly* when exposed to light, and silver is deposited.

  $Ag^+ + light + e^- \rightarrow Ag(s)$

- In the body, the process of oxygen bonding to the iron in haemoglobin as blood passes through the lungs is 'instantaneous', whereas the various processes of digestion may take several hours.

  Most reactions that you will study happen in a matter of minutes; some of these will be used as examples to show how the rate of a reaction might be quantitatively measured .

  In Unit 5.3 the ways in which reactions can be made to proceed more or less rapidly are discussed.

THE UNITS USED IN THE MEASUREMENT OF REACTION RATES

The speed of a car is measured in kilometres per hour ($km\ h^{-1}$) or miles per hour (mph); the pulse rate is a measure of how many times the heart beats in a minute. In a chemical reaction, as the products are formed so the reactants are used up. We say that the **concentrations** of the reactants and products change. **Concentration** is the amount of substance (worked out in moles) in a certain volume (measured in cubic decimetres or litres).

One cubic decimetre = one thousand cubic centimetres = one litre
$1\ dm^3 = 1000\ cm^3 = 1\ l$
Concentration is measured in **moles per cubic decimetre**
i.e. **$mol\ dm^{-3}$**

- The **rate of a chemical reaction** is the time taken, in seconds, for a specified amount of reactant to be used up or product to be formed.
- The **rate of a chemical reaction** then, is the time taken **in seconds** for a measured **change in the concentration** of a reactant or product.

Rate of Reaction = mol dm^{-3}/second
which can be written as **mol dm^{-3} s^{-1}**

Any property of the system that changes during the course of the reaction, such as colour or opacity of solution, or mass loss, or volume changes in gaseous systems, can be used to monitor the rate. In practice, chemists will choose the simplest available method.

5.2.2 Measuring the Rates of Chemical Reactions

As a chemical reaction proceeds the number of particles (moles) of the reactants decreases and the number of particles of the products increases. To measure the rate of a chemical reaction, we must find out how fast the particles of reactant are disappearing, or how fast the product is being made. Fortunately we do not have to try to count the actual number of individual particles but can monitor a related property, such as the change in concentration of a solution or the volume of gas evolved.

THE REACTION BETWEEN CALCIUM CARBONATE AND HYDROCHLORIC ACID

$$CaCO_3(s) + 2HCl(aq) \rightarrow CaCl_2(aq) + CO_2(g) + H_2O(l)$$

In this reaction we could monitor how fast carbon dioxide gas or calcium chloride was being made, or how quickly calcium carbonate or hydrochloric acid was being used up. The simplest method is to track the amount of carbon dioxide evolved, which can be done in two ways.

a) As the reaction proceeds an increasing volume of carbon dioxide gas is evolved, which is easily collected in a gas syringe or over water in a burette. The volume of gas collected is noted every few seconds, as shown in Figure 5.2.1.

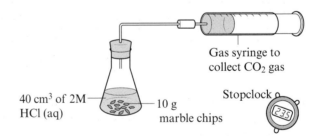

Gas syringe to collect CO$_2$ gas

Figure 5.2.1
Experimental apparatus
to measure the volume of
gas evolved

40 cm^3 of 2M HCl (aq)

10 g marble chips

Stopclock

b) The mass of the reacting system in the conical flask decreases as the calcium carbonate reacts to form the products and gas is evolved. So it is quite straightforward to monitor the change in mass of the system with time, as shown in Figure 5.2.2.

Figure 5.2.3 shows the results of these methods and how the rate of a reaction between calcium carbonate and hydrochloric acid varies with time. The graphs have different shapes because one charts the change in formation of carbon dioxide, *i.e.* the products, and the other shows how the reaction changes with relation to the reactants, *i.e.* how fast the system loses mass as the calcium carbonate and hydrochloric acid react to form carbon dioxide.

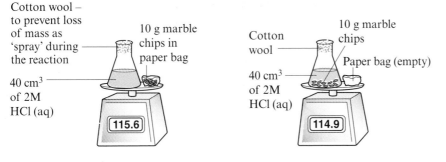

Figure 5.2.2
Monitoring the change in mass
with time

Both of the curves in Figure 5.2.3 show that as the reaction proceeds it begins to slow down, *i.e.* the rate decreases. Eventually, the reaction must stop because one of the reactants has been used up and so no more carbon dioxide will be produced – shown by the way that both curves 'level off'. The rate of loss in mass of the system is directly related to the speed at which the reactants are being used up and, since the *volume* of the acid remains constant, can be taken as a measure of the way that the *concentration* of the acid is changing.

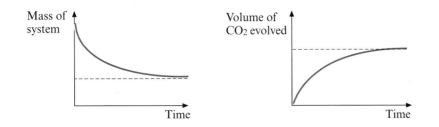

Figure 5.2.3
Both graphs show the progress
of the same reaction

Sample results: measuring rates of reactions

The results in Table 5.1.1 were collected from two sets of experiments set up like those in Figures 5.2.1 and 5.2.2 above. The table shows the results obtained when the loss in mass is recorded when 15 g of calcium carbonate as marble chips, react with 50 cm^3 of 2 mol dm^{-3} hydrochloric acid.

Table 5.1.1 *Table of results obtained in the study of the reaction between dilute hydrochloric acid and calcium carbonate, as 'large' marble chips*

Time/s	0	20	40	60	90	120	150	180	210	240	300
Loss in mass/g	0	0.30	0.60	0.94	1.22	1.42	1.60	1.72	1.80	1.82	1.84

When the data are plotted we get the curve shown in Figure 5.2.4. The loss in mass of the system is equal to the mass of the carbon dioxide formed because the gas escapes through the plug of cotton wool in the neck of the flask. So the graph actually shows the formation of carbon dioxide with time.

MEASURING THE RATE OF REACTION

The **rate of the reaction**, at a specific time, can be worked out by finding the slope or *gradient* of a tangent to the curve at that time.

$$\text{The rate of reaction} = \frac{\text{change in amount of products}}{\text{time taken for change to occur}}$$

From the graph:

Let the time taken to evolve 240 cm³ CO_2 at room temperature be t_1,

$$t_1 = 120 \text{ s}$$

Let the time taken to evolve 240 cm³ CO_2 at 30 °C be t_2, $t_2 = 62 \text{ s}$

In both cases the time has been recorded for the formation of the same amount of product, so we can use the approximation:

- Rate of reaction = $1/t$
- At room temperature, it takes 120 s for 0.01 mol of carbon dioxide gas to be evolved, so the average rate of formation is $1/120 = 0.0083 \text{ s}^{-1}$.
- At 30 °C it takes 62 seconds for 0.01 mol of carbon dioxide gas to be evolved, so the average rate of formation is $1/62 = 0.016 \text{ s}^{-1}$.

THE REACTION BETWEEN SODIUM THIOSULFATE AND HYDROCHLORIC ACID

In this reaction the fine suspension of minute sulfur particles increases as the time elapses making the liquid cloudy. So, the faster the reaction proceeds, the faster the cloudiness appears, and this can be used as a measure of the rate of reaction. When the solutions are quickly mixed together, the time taken for the deposit of sulfur to obscure a pencilled cross, drawn on paper beneath the conical flask, is inversely proportional to the rate of the reaction (see Figure 5.2.6).

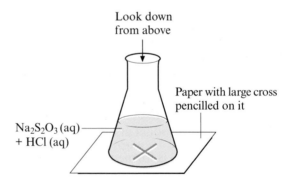

Figure 5.2.6
Experimental apparatus to study the formation of sulfur

$$Na_2S_2O_3(aq) + 2HCl(aq) \rightarrow 2NaCl(aq) + H_2O(l) + S(s) + SO_2(g)$$

It is possible to investigate the reaction for a variety of different conditions. In order to make fair comparisons, it is imperative that only one factor is changed each time.

Example:
To study the effect, on the rate, of changing the concentration of sodium thiosulfate solution, we must keep constant:

a) the concentration of hydrochloric acid;
b) the overall volume of solutions, so that the depth of liquid above the cross is the same;
c) the temperature.

5.2.3 Different Methods used to Monitor the Progress of Some Reactions

Not all reactions involve such obvious and easily followed changes as those used in the above examples. The method used depends on the type of reaction that is occurring. Here are some other methods that might be used:

- **Titration** – for reactions such as esterification or the hydrolysis of an ester, a sample may be removed from the reacting system at regular time intervals and, for example, the amount of acid present can be found by carrying out a titration against an alkali of known concentration.
- **Colour changes** – many chemicals are coloured, and the intensity of the colour of their solutions will vary with concentration; so as the reaction proceeds and products are formed, colour intensity can be monitored using a colorimeter to determine the rate.
- **Volume or pressure changes** – reactions between gases can be monitored by measuring changes in volume or pressure. In practice, the latter method is most usual, as it is difficult to change the volume of the apparatus.

  e.g. $2NO(g) + O_2(g) \rightarrow 2NO_2(g)$

  You will notice that there is a reduction in the number of moles of gas present – so if the volume of the reacting vessel is kept the same there will be a drop in pressure.
- **Rotation of plane polarised light** – some compounds can rotate the plane of polarised light (see Unit 6.4). For example, when sucrose reacts with water it forms glucose and fructose which have different structures and rotate the light to different degrees. This change can be monitored by using a polarimeter.
- **Conductivity changes** – the conductivity of an ionic solution is proportional to the number of ions present. Many reactions involve ionic solutions, and the changes in concentration of the ions present can be monitored *via* changes in conductivity.

Answers to diagnostic test

If you score less than 80%, then work through the text and re-test yourself at the end by using this same test. If you still have a low score then re-work the topic at a later date.

1 a) An experimental setup is given in the following diagram.

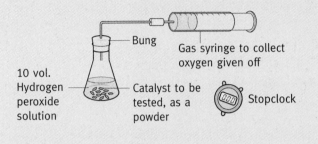

- Use a fixed volume of hydrogen peroxide solution of known concentration. **(1)**
- Carry out a 'control' reaction at room temperature and measure the time for a specific volume of oxygen to be evolved, or see how much oxygen is evolved in a certain time period. **(2)**
- Keep the concentration of the peroxide solution, the temperature, and all other factors constant, but repeat the experiment using manganese(IV) oxide, copper(II) oxide, iron(III) oxide, *etc.* as catalysts. **(2)**

b) See Figure 5.2.6 for suitable apparatus. **(1)**
- Use fixed volumes, of known concentration, of sodium thiosulfate solution and hydrochloric acid: time how long it is before the cloudiness of the solution obscures the cross. **(2)**
- Keep the temperature, overall volume and the concentration of the sodium thiosulfate solution the same as before, and dilute the acid in a controlled way. Run the experiment again using the different concentrations of acid. **(2)**

2 a) mol dm^{-3} s^{-1} **(2)**

b) Reactions slow down as the reacting substances are used up – and stop when one or both reactants are completely used up. **(2)**

3 The answers to (a) and (b) are shown in the graph below.

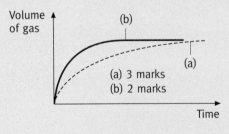

(5)

4 $Mg(s) + H_2SO_4(aq) \rightarrow MgSO_4(aq) + H_2(g)$

48 cm³ of hydrogen forms in 30 s

a) Rate of formation = 48/30 = 1.6 cm³ s⁻¹ **(1)**

b) 1 mole   hydrogen occupies   24 dm³ at STP
                                          (= 24 000 cm³)

  1/24 000 mole   hydrogen occupies   1 cm³ at STP **(1)**

  ∴ $\dfrac{1.6 \text{ cm}^3 \text{ mole}}{24\,000}$ hydrogen occupies   1.6 cm³ at STP **(1)**

  0.000 067 mole   hydrogen occupies   1.6 cm³ at STP **(1)**
  6.7 × 10⁻⁵ mole   hydrogen occupies   1.6 cm³ at STP **(1)**

  ∴ rate of reaction = 6.7 × 10⁻⁵ mol s⁻¹

  So, hydrogen is being formed at the rate of
  6.7 × 10⁻⁵ mol s⁻¹ **(1)**

25 Marks   **(80% = 20)**

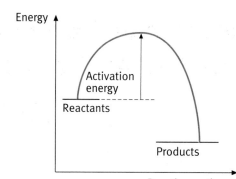

Figure 5.3.1
Energy diagram

15 Marks   (80% = 12)

5.3.1 Introduction to the Effect of Temperature on Reactions

The rates of chemical reactions are altered by changing the conditions; for instance, milk left out in a warm kitchen goes sour much more quickly than when kept in the refrigerator. Food that has been dehydrated stays safely edible longer than the original product.

These are both examples of common chemical reactions which are affected by changes in the physical conditions.

Chemical reactions can be affected by various factors such as:

- changes in temperature;
- changes in the concentration of reactants in solution;
- changes in the surface area of solid reactants;
- changes in pressure in gaseous systems;
- catalysts;
- light.

This unit looks at the influence of some of these factors on some common reactions and uses theoretical ideas to explain the behaviour.

THE REACTION BETWEEN CALCIUM CARBONATE AND HYDROCHLORIC ACID

This reaction may again be used to illustrate the way in which this may be measured. The same apparatus may be used as in Figure 5.2.1 in Unit 5.2, but the flask and acid would have to be heated (in a constant temperature water bath for greatest accuracy) prior to adding the calcium carbonate. In this experiment, the volume and concentration of acid and the particle size and mass of the calcium carbonate *must all be kept constant*.

Figure 5.3.2 hows that carbon dioxide gas is given off faster at higher temperatures (the curve is initially steeper).

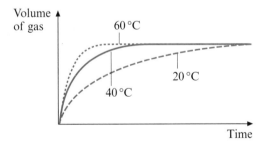

Figure 5.3.2
Volume of gas evolved at different temperatures

THE REACTION BETWEEN SODIUM THIOSULFATE AND HYDROCHLORIC ACID

For the method and apparatus, see page 222 in Unit 5.2.

The results in Table 5.3.1 show the effect of changing the temperature and measuring the time for enough sulfur to be formed to obscure the pencilled cross.

Table 5.3.1 *Results from an investigation into the effect of heat on the decomposition of sodium thiosulfate solution*

Time (s)	Temperature (°C)	Rate of reaction given by 1/t (s⁻¹)
288	10	0.0035
170	20	0.0058
90	30	0.0110
43	40	0.0233
20	50	0.0500
11	60	0.0910

Figure 5.3.3 shows that the rate of reaction increases sharply as the temperature is increased. The average rate at 10 °C is 0.0035 s⁻¹, which is approximately half the average rate at 20 °C, 0.0058 s⁻¹.

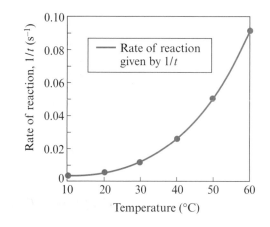

Figure 5.3.3
Rate of reaction as a function of temperature

Experimentation shows us that a temperature rise has a big effect on the rate.
• For many reactions **the rate doubles for each 10 °C rise in temperature**.

• *So, how much faster would a reaction be at 80 °C, compared with its rate at 20 °C?*
• *It will be about 64 times faster!*

Why does temperature have such a big effect on reaction rate?

Raising the temperature increases the **kinetic energy** of the reacting particles, so there will be more collisions between the particles; more important still, the collisions have more energy.
The relationship between kinetic energy (KE) and speed of particles is given by the equation,

$$KE = \tfrac{1}{2}mv^2$$

and since the mass of the particles is constant for each reaction,

$$KE \propto v^2$$

If, say, $v = 4$ $KE \propto 4^2 = 16$

If, now, $v = 8$ $KE \propto 8^2 = 64$

• Doubling the speed of the particles quadruples the kinetic energy.
• So relatively small increases in temperature have a large effect on KE and thus on the rate of reaction.

5.3.2 The Effect of Changing the Concentration of Reactants in Solution

If 5 g of salt are dissolved in a bucket of water and a test tube of water, the solution in the latter is more concentrated – there is the same amount of salt particles but in a much smaller volume – the particles are closer together – the solution is *more concentrated*.

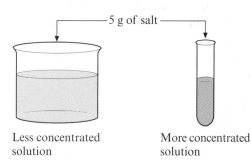

Less concentrated solution

More concentrated solution

Figure 5.3.4
The same amount of salt in different volumes of water

Consider again the reaction between calcium carbonate, as marble chips, and hydrochloric acid.

$$CaCO_3(s) + 2HCl(aq) \rightarrow CaCl_2(aq) + CO_2(g) + H_2O(l)$$

The reaction temperature and the mass and particle size of the calcium carbonate are kept the same, but different concentrations of acid are used.

e.g. 100 cm³ of 0.25 mol dm⁻³ hydrochloric acid contains the same amount of acid (0.025 moles) as 50 cm³ of 0.50 mol dm⁻³ acid and 25 cm³ of 1.0 mol dm⁻³ acid, *but* the *concentrations* of the acid solutions are different! (For more information about moles, see Module 4.)

NOTE: It is important to use sufficient volume of acid in each case to cover the marble chips.

NOTE: An excess of marble is always used to ensure a 'fair test'; *i.e.* the acid will be totally used up during the course of the reaction.

Figure 5.3.5 shows the results from using three different concentrations of acid (all containing the same number of moles) and measuring the volume of carbon dioxide gas evolved.

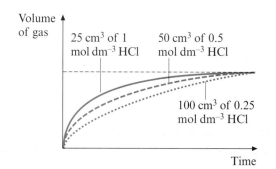

Volume of gas

25 cm³ of 1 mol dm⁻³ HCl

50 cm³ of 0.5 mol dm⁻³ HCl

100 cm³ of 0.25 mol dm⁻³ HCl

Time

Figure 5.3.5
Volume of gas evolved using different concentrations of acid

At higher concentrations the gradient of the curve is initially steeper, showing that *the higher the concentration of the acid then the faster the reaction rate*.

Why?

In a more concentrated solution there are more particles which are also closer together, so the chances of collisions are increased, so the rate will increase.

5.3.3 The Effects of Pressure Change on the Rate of a Gaseous Reaction

When the pressure on a gas is increased, if the gas cannot escape, the particles are forced closer together (Figure 5.3.6). The effect is similar to that of increasing the concentration of substances in solution – at higher pressures there will be more particles squeezed into the same space, so there is a greater likelihood of collisions and the reaction proceeds more quickly. This is put to effective use in the production of ammonia by the Haber Process, $N_2(g) + 3H_2(g) \rightleftharpoons 2NH_3(g)$, (see Unit 5.4).

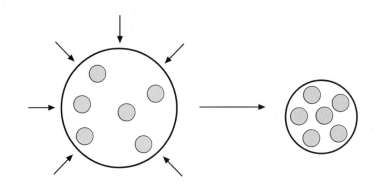

Figure 5.3.6
If pressure is applied to the gas, the particles are pushed closer together

5.3.4 The Effect on the Rate of Reaction of the Surface Area of Solid Reactants

A good demonstration of the effect of particle size is to blow a drinking straw full of a finely-divided combustible powder – such as custard powder – through the flame of a roaring Bunsen. Make sure no-one is standing too close on the other side!

When fine powders spread out into the air, they expose a vast surface area to oxygen molecules in the air. There have been devastating explosions in flour mills when finely powdered flour has mixed with air. The mixture explodes if there is the slightest spark, say from a nail in a shoe striking the floor. The BANG that is heard is the result of the extremely fast reaction, as it produces products so violently that the molecules fly apart faster than the speed of sound – they produce a shock wave that breaks the 'sound barrier'. Mixtures of coal dust and air explode similarly, and have caused thousands of deaths in the coal-mining industry.

Consequently, manufacturers handling such fine powders must take special precautions against fires and explosions.

It seems then, that particle size affects the rate of chemical reactions. This can be demonstrated in the laboratory by monitoring a reaction in which the solid involved in a chemical reaction is subdivided into smaller and smaller pieces. But first, let us do some maths!

Consider a one centimetre cube:

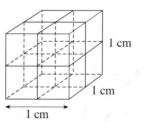

Figure 5.3.7
Smaller particles have a greater total surface area

• Surface area = 6 cm²

If we slice the cube in half horizontally and vertically, we now have eight cubes, each one with sides ½ cm long.

• Each ½ cm cube has a surface area of 6 × (½ × ½) cm² = 1½ cm²
• The total surface area of the substance is 12 cm² [8 × 6 × (½ × ½)]

If we then slice the ½ cm cubes in half horizontally and vertically, we will have 64 cubes, each of side ¼ cm.

• Total surface area of the substance is now 24 cm² [64 × 6 × (½ × ½)]

Each time that the dimensions of the particles are *halved*, the total surface area *doubles* (for the same mass of solid).

Can you imagine the total area of the original cube if it were crumbled to micron-size particles?

In the reaction between dilute hydrochloric acid and marble chips (calcium carbonate),

$$CaCO_3(s) + 2HCl(aq) \rightarrow CaCl_2(aq) + CO_2(g) + H_2O(l)$$

the marble is easily broken into smaller pieces or ground into powder using a pestle and mortar. Figure 5.3.8 shows the effect on the rate of reaction.

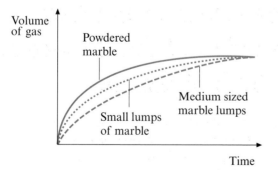

Figure 5.3.8
Volume of gas evolved using different sized marble particles

When a solid reacts with a liquid or a gas the collisions between the particles must occur at the surface of the solid. If the solid is subdivided (broken up) there will be more of its surface exposed to the other reactants – so there will be more collisions per second at the surface, resulting in an increase in the reaction rate.

5.3.5 The Effects of Catalysts on the Rate of Reaction

An ideal catalyst will increase the speed of a reaction without itself being destroyed or used up during the reaction.

When a gas jar of hydrogen is opened in air, nothing appears to happen, but if some mineral wool coated in finely divided, platinum black powder is held at the mouth of the gas jar, then the '*squeaky*' pop of the mixture exploding will be heard *even without putting a flame to the mixture*:

$$2H_2(g) + O_2(g) \xrightarrow{\text{Pt}} 2H_2O(l)$$

The platinum has acted as a **catalyst** – it has speeded up the reaction. Many chemical reactions are speeded up by catalysts; some are important in industry, and others occur in every living cell.

WHAT DO WE KNOW ABOUT CATALYSTS?

- Catalysts dramatically speed up some reactions, so that products are formed or equilibrium is reached more quickly (see Unit 5.4).
- Catalysts are not *chemically* changed by the reactions (although physically they may look different) and they can **always** be recovered afterwards.
- Catalysts can change the **rate** of an equilibrium reaction, by allowing equilibrium to be reached more quickly, but not the **extent** (how far it will go) – *i.e. catalysts do not make reactions form more products* (see Unit 5.4).
- Some solid catalysts are made more effective by increasing their surface area and sometimes by using high temperatures, although there is often an optimum temperature at which the catalyst is most effective.
- Catalysts tend to be specific; just because they alter the rate of one reaction does not mean that they will affect another. Different reactions are catalysed by different substances.
- Some substances that look quite similar can have very different effects on the same reaction. For example, Figure 5.3.9 shows that powdered black manganese(IV) oxide speeds up the decomposition of hydrogen peroxide solution, whereas black copper(II) oxide powder is not so effective:

$$2H_2O_2(aq) \xrightarrow[\text{or CuO}]{\text{MnO}_2} 2H_2O(l) + O_2(g)$$

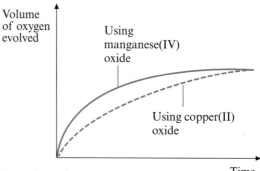

Figure 5.3.9
H_2O_2 is decomposed more effectively using MnO_2 as the catalyst

THE PHYSICAL STATE OF CATALYSTS

Heterogeneous catalysts

When the catalyst used is in a *different physical state* to the reactants it is called a *heterogeneous catalyst*. Solids are often heterogeneous catalysts; for example, platinum powder in the reaction between oxygen and hydrogen gases, and powdered manganese(IV) oxide in the decomposition of aqueous hydrogen peroxide. Reactions involving solid catalysts occur at the surface of the solid, so they are often most effective when the catalyst is ground into a powder or used as a gauze; both methods increase the surface area available for reaction. It is thought that catalysts work by adsorbing[1] reacting materials at their surface, forming a temporary bond: this weakens other bonds in the reactant and improves the opportunity for reaction (see page 247). Solid catalysts are easily 'poisoned' by impurities in the reactants, if these are also readily adsorbed by the catalyst, because they clog the catalyst's surface.

The Active Site

Many metals or metal oxides act as catalysts for gaseous reactions. Consider the reaction between ethene, C_2H_4, and hydrogen. The reaction is catalysed by nickel. At certain sites on the surface of the nickel the arrangement of atoms encourages bond breaking. These places are called **active sites**. Active sites are usually associated with some irregularities in the surface of the substance. Many metal catalysts have small amounts of impurities (known as 'promoters') deliberately added to them to 'disturb' the regular metal lattice, and create 'bumps' which are, on average, the right distance apart to encourage adsorption of reacting molecules (Figure 5.3.10). For example, in the Haber process the finely divided iron used as a catalyst may have potassium oxide added to act as a 'promoter'.

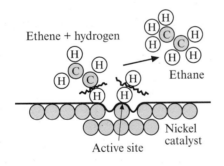

Figure 5.3.10
Active sites on nickel speed up the reaction between ethene and hydrogen

Homogenous catalysts

When the reactants and the catalyst are in the *same physical state* the catalyst is called a *homogeneous* catalyst – *e.g.* concentrated sulfuric acid speeds up the reaction between ethanoic acid and ethanol to form the ester, ethyl ethanoate. Manganese(IV) oxide catalyses the thermal decomposition of potassium chlorate(V) to give oxygen.

$$2KClO_3(s) \xrightarrow{\text{heat; } MnO_2} 2KCl(s) + 3O_2(g)$$

---

[1] Adsorption occurs on the surface of a solid, whereas absorption occurs within the substance.

HOW DO CATALYSTS WORK?

Catalysts increase the rate of a reaction by allowing reactions to proceed *via* a different pathway which has a lower **Activation Energy** (see Unit 5.1). The effect is that, even without heating, more particles have a kinetic energy sufficient to cause a reaction (Figure 5.3.11).

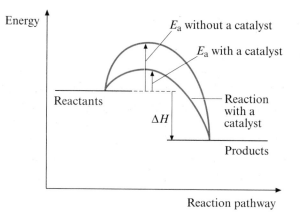

Figure 5.3.11
A catalyst provides a lower energy pathway for the reaction

Although a catalyst lowers the activation energy (E_a), the overall (net) heat or energy change for the reaction (ΔH) is unaltered.

'NEGATIVE' CATALYSTS

Some substances are capable of *slowing down* reactions. These are sometimes called *negative catalysts* or *inhibitors*. There is however, considerable debate as to whether or not they are *strictly* catalysts, because they usually work by *combining chemically* with one of the components in a reaction. For example, tetraethyl-lead(IV), $Pb(CH_3CH_2)_4$, known as 'anti-knock', is added to petrol. Petroleum mixtures that are rich in straight chain alkanes (see Module 6) ignite very easily and explode rapidly, causing 'knocking' and inefficient combustion. When tetraethyl-lead(IV) burns it forms small particles of lead(II) oxide which combine with free radicals, produced when petrol burns, and thus slows down the reaction, making it smoother.

Styrene, $C_6H_5CH{=}CH_2$, is the monomer used to make polystyrene; it is readily activated by light if left in a clear glass bottle, forming free radicals that rapidly polymerise. To prevent unwanted polymerisation, styrene is kept in solution with a second substance, which acts as an inhibitor by reacting with free radicals more efficiently than the monomer.

Perhaps the most common use of inhibitors is as preservatives for the food and cosmetics industry; on labels they are often called antioxidants, *i.e.* they prevent the oxidation of certain substances in the food or cosmetic, keeping it 'fresh' for longer. For example, sulfur dioxide preserves dried fruit, peas and beans.

ENZYMES AS CATALYSTS

Some enzymes are not only extraordinarily specific in their reactions, but also almost unbelievably fast. Some have been shown to carry out many thousands of reaction cycles in a second!

The body is the most efficient set of complementary reactions known, mainly due to the effects of special catalysts called enzymes. Most reactions occurring in the body – and indeed in all living cells – are catalysed by enzymes. Enzymes are very specific catalysts; each enzyme affects only one reaction and a single cell may contain many different enzymes.

When glucose releases energy in the muscles, a small amount of hydrogen peroxide is sometimes formed as a by-product, which can damage the body. The blood contains an enzyme called catalase which effectively speeds up the decomposition of the hydrogen peroxide into water and oxygen, rendering it harmless:

$$2H_2O_2(aq) \xrightarrow{\text{catalase}} O_2(g) + 2H_2O(l)$$

Catalase is a more effective catalyst on this reaction than manganese(IV) oxide. This can be shown by adding a piece of liver or liquidised celery (which both contain catalase) to a solution of hydrogen peroxide and observing that oxygen gas is evolved instantly and dramatically.

People have made use of enzymes in reactions for thousands of years, for instance in fermentation and cheese-making processes, but it is only recently that their identity and the importance of their role has been recognised and understood.

What is an enzyme and how does it work?

Enzymes are long-chain **proteins** that have a complicated structure and normally have only one particular site at which a reaction can take place. X-Ray diffraction results have shown that this region, called the active site, has a definite shape for each enzyme, and therefore only molecules with a similar complementary shape will fit into it – rather like keys in a lock. For example, only hydrogen peroxide molecules will fit the active site of catalase. Those molecules that do fit the active site are called substrate molecules. Thus, enzymes are *much more* specific catalysts than inorganic substances such as manganese(IV) oxide.

It can be helpful to picture the action between enzyme and substrate as being like fitting together two pieces of jigsaw as in Figure 5.3.12. The stages are:

Since enzymes work by a **surface** lock and key effect they might be thought of as heterogeneous catalysts. However, they also react in solution in body fluids which suggests homogeneous catalysis. In reality, it is hard to put enzymes into either category.

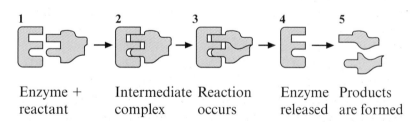

1	2	3	4	5
Enzyme + reactant	Intermediate complex	Reaction occurs	Enzyme released	Products are formed

Figure 5.3.12
The 'lock and key' effect of enzyme and substrate

Like most other catalysts, enzymes can be *poisoned* if the active sites become clogged by an unwanted molecule. Inorganic metal catalysts are easily spoiled by hydrogen sulfide (H_2S) or arsenic, whereas enzymes are particularly vulnerable to changes in hydrogen ion concentrations (pH) and temperature. Enzymes work best at temperatures around 37 °C (body temperature) and at a pH around 7 (neutral), although the digestive enzymes in the stomach work at a much lower pH.

Changes in pH and temperature inhibit or prevent the reaction at the enzyme's active site. The peptide chain is constructed from amino acids (see Module 7), which can gain or lose electrons; consequently enzymes are ruined by conditions that are too alkaline or acidic [Figure 5.3.13(a)]. Even very small changes in pH around the active site can prevent the substrate from entering or leaving.

Increases in temperature alter the protein structure (it is *denatured*) at the active site so that the enzyme can no longer hold the substrate. This is because much of the molecular structure of the enzyme depends on

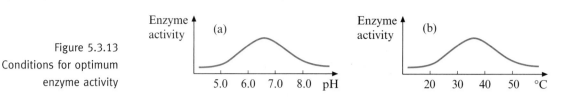

Figure 5.3.13
Conditions for optimum
enzyme activity

hydrogen bonding (see Module 2), which is easily disrupted by increased temperatures [Figure 5.3.13(b)].

Despite their extreme sensitivity, and sometimes high cost, enzymes are being increasingly used in industrial processes, especially the food industries.

* **Fermentation** is catalysed by the enzymes in yeast.
* Making **cheese**, proving **bread** dough, brewing **beer** and **tenderising meat** by hanging it are all processes dependent on enzymes.

Enzymes can be used to break down certain molecules, literally 'pre-digesting' them, making foods more palatable or easier to digest. This process has been used:

* in making baby foods;
* in tenderising meat;
* in forming soft-centred sweets.

Enzymes are also used in the manufacture of some soap powders, where the enzyme action breaks down (digests) molecules of oils, dirt, *etc.*, and in the treatment of leather.

Some common enzymes and their functions are shown in Table 5.3.2.

Table 5.3.2 *Some enzymes and their functions*

Enzyme	Function	Use
Amylase	Converts 'starchy' complex carbohydrates to sugars	Digestion
Catalase	Converts hydrogen peroxide to water and oxygen	Removes H_2O_2, formed as a waste product of respiration, from the blood
Protease	Breaks down protein molecules	Biological wash powders – for removal of blood and gravy stains, *etc.*
Invertase	Converts sucrose to fructose and glucose	Making confectionery – especially 'soft-centred' sweets
Lysozyme	Breaks down the polysaccharide walls of bacteria	Helps the body to combat infection
Papain	Breaks down long protein chains	Used to tenderise meat
Urease	Converts urea to ammonia and carbon dioxide	Breakdown of urea

THE KINETICS OF ENZYME REACTIONS

In reactions that are catalysed by enzymes, experiments show that the rate increases with the concentration of the enzyme. The rate also increases if the concentration of the substrate is increased – but only up to a point. The reaction rate reaches a limit because eventually the active site becomes saturated with substrate molecules. When this happens, adding more substrate will have no effect on the rate, because all the active sites of the enzyme have substrate molecules attached to them (Figure 5.3.14).

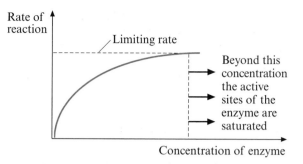

Figure 5.3.14
Effectiveness of an enzyme depends on the concentration of the substrate

ENZYME INHIBITORS

It has been noted that certain substances slow down and even stop enzyme-controlled reactions; they are called inhibitors. Competitive inhibition occurs when a molecule has a similar molecular configuration to the substrate: it therefore competes with the normal substrate for the active site and may slow down the reaction. The degree of inhibition depends on the relative concentrations of the substrate and inhibitor. Non-competitive inhibition occurs when the inhibitor attaches itself permanently to the active site: the extent of inhibition depends entirely on the inhibitor concentration and cannot be altered by changing the amount of substrate present. Arsenic and heavy metals such as mercury and silver are toxic because they are inhibitors (non-competitive). Nerve gas, developed during the Second World War, is another example of an inhibitor; it combines competitively with the enzyme cholinesterase and slows down the transmission of nerve impulses from one cell to another.

Enzyme inhibitors are not always 'bad news'; moment-to-moment variations in the rate of cellular metabolism are caused by the control of enzyme action by inhibtors that occur naturally in the body.

Question

Catalysts A and B both catalyse the decomposition of hydrogen peroxide to give oxygen gas. The data in Table 5.3.3 were obtained when the two catalysts were used.

Table 5.3.3 *Results from an investigation of the effect of different catalysts on the decomposition of hydrogen peroxide solution*

Time (min)	0	5	10	15	20	25	30	35
Volume of O_2 gas (cm^3) with catalyst A	0	4	8	12	16	17	18	18
Volume of O_2 gas (cm^3) with catalyst B	0	5	10	15	16.5	18	18	18

a) Using the *same pair of axes*, plot graphs of both sets of results.
b) Which is the better catalyst, and what is the evidence for your decision?
c) Why does the volume of oxygen collected stop at 18 cm^3 in both sets of results?
d) Sketch a line on your graph to show the shape of the graph you would obtain for a non-catalysed reaction.
e) What factors had to be kept the same when carrying out the experiments to compare the two catalysts, and why?

Answers to diagnostic test

If you score less than 80%, then work through the text and re-test yourself at the end by using this same test. If you still have a low score then re-work the unit at a later date.

1 Two chemicals will only react if the particles collide and have sufficient kinetic energy to overcome the energy barrier (activation energy).   **(2)**

2 a) Heating the reactants gives particles <u>greater kinetic energy</u> and so there <u>is more chance</u> of <u>high energy collisions</u> that are successful.   **(3)**

    b) More <u>chances of collisions</u> because there are <u>more particles per unit volume</u>.   **(2)**

3 a) A catalyst is a substance that can <u>speed up</u> a chemical reaction and can be <u>recovered chemically unchanged</u> at the end of the reaction.   **(2)**

    b) For a catalysed reaction the energy diagram changes to:

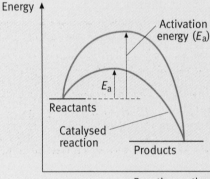

(2)

4 Although there might be some exceptions, enzymes are generally constructed in a way very similar to those in the body, so most enzymes function optimally around <u>pH about 7</u> and <u>at a temperature of 37 °C</u>.   **(2)**

5 Heat means money! Reactions needing <u>high temperatures are costly</u>, involving large amounts of fuel and possibly cooling systems also to reduce product temperatures. A catalyst <u>speeds up the reaction and is less costly</u>.   **(2)**

15 Marks   (80% = 12)

Unit 5.4
Reversible Reactions and Chemical Equilibrium

Starter check

What you need to know before you start this unit.

You should be able to interpret chemical equations, and appreciate that they represent definite quantities of materials reacting together. You should be familiar with the concepts of exothermic and endothermic reactions.

Aims **By the end of this unit you should understand and be able to:**

- Explain the effect of different factors on the equilibrium state of various reactions.
- Appreciate the implications of equilibrium states in industrial processes.

Diagnostic test

Try this test at the start of the unit. If you score more than 80%, then use this unit as a revision for yourself and scan through the text. If you score less than 80%, then work through the text and re-test yourself at the end by using this same test.

The answers are at the end of the unit.

1 What is Le Chatelier's principle? (1)

2 When an equal quantity of iron(II) sulfate solution is added to silver nitrate solution of the same concentration, an equilibrium mixture is formed and a precipitate of silver is seen. The ionic equation is:

$Fe^{2+}(aq) + Ag^+(aq) \rightleftharpoons Ag(s) + Fe^{3+}(aq)$

a) What would happen if the concentration of the iron(II) sulfate solution was increased? (2)

b) What would you expect to see if iron(III) sulfate solution was added to the equilibrium mixture? (2)

3 Using an example for each, explain the difference between reactions involving:

a) thermal dissociation; (2)

b) thermal decomposition. (2)

4 In the Haber process, ammonia is made in an exothermic reaction between hydrogen and nitrogen gases:

$$N_2(g) + 3H_2(g) \rightleftharpoons 2NH_3(g) \quad \Delta H = -92 \text{ kJ mol}^{-1}$$

Explain, with reasons, what the effect would be on the proportion of ammonia in the equilibrium mixture if:

a) the temperature was increased; (3)

b) the pressure on the system was increased. (3)

5 a) Write an equation for the reaction between ethanol and ethanoic acid. (2)

b) Write an expression for the equilibrium constant for this reaction. (3)

20 Marks   (80% = 16)

5.4.1 Reversible and Non-reversible Reactions – Introduction

Boiling an egg and burning wood or paper in air are similar processes in that it is impossible to get back what you started with. These are both one-way, or irreversible, reactions. But many processes and reactions can be reversed by changing the conditions.

Most *physical* changes are easily reversed; for instance jelly sets on cooling but becomes a liquid again if warmed. Water, as ice, can be heated, becoming first liquid water and then steam, which can then be condensed back to water and frozen to re-form ice:

$$H_2O(s) \rightarrow H_2O(l) \rightarrow H_2O(g)$$

$$H_2O(g) \rightarrow H_2O(l) \rightarrow H_2O(s)$$

These equations can be put together to show that, overall, the changes can operate in either direction:

$$H_2O(s) \rightleftharpoons H_2O(l) \rightleftharpoons H_2O(g)$$

When a compound is split up by the action of heat, the reaction may be reversible or irreversible. When the reaction is non-reversible it is called **thermal decomposition**:

a) Solid potassium chlorate decomposes to produce solid potassium chloride and oxygen gas: the two products cannot be made to recombine:

$$2KClO_3(s) \rightarrow 2KCl(s) + 3O_2(g)$$

b) Ammonium nitrate decomposes to give the gas of dinitrogen oxide and water:

$$NH_4NO_3(s) \rightarrow N_2O(g) + 2H_2O(g)$$

When a reaction can be reversed by altering the temperature, the process is called **thermal dissociation**:

a) Ammonium chloride is one of the few compounds that sublimes – it turns directly from a solid to a gas when heated:

$$NH_4Cl(s) \rightleftharpoons NH_3(g) + HCl(g) \quad \Delta H = +177.2 \text{ kJ mol}^{-1}$$

On cooling, so long as none of the gases have been allowed to escape, the two products recombine to form solid ammonium chloride again.

b) If limestone, natural calcium carbonate, is heated strongly in a sealed container, it is partly split into calcium oxide (quicklime) and carbon dioxide and the reaction is reversible; but if the gas is allowed to escape then decomposition is completed:

Forward $\longrightarrow$
$$CaCO_3(s) \rightleftharpoons CaO(s) + CO_2(g)$$
$\longleftarrow$ Backward

- Physical changes are readily reversed by a change of the physical conditions.
- Chemical changes (reactions) involve the formation of new substances. They cannot be reversed by any simple means.

Some reactions may be reversed by changing the chemical conditions; for instance hot iron reacts with steam to form magnetic iron oxide and hydrogen:

$$3Fe(s) + 4H_2O(g) \rightleftharpoons Fe_3O_4(s) + 4H_2(g)$$

The reverse reaction may be carried out by reducing hot iron oxide with hydrogen:

$$Fe_3O_4(s) + 4H_2(g) \rightleftharpoons 3Fe(s) + 4H_2O(g)$$

5.4.2 Systems at Equilibrium

PHYSICAL EQUILIBRIUM

When sugar is added to a water it begins to dissolve. The structure of the crystal begins to break down and individual particles are able to move from the solid into the liquid forming a solution. At the same time a few particles of sugar will 'fall out of solution' and be deposited back onto the surface of the solid. Eventually there comes a point when the surface area of the solid is so small that fewer and fewer particles are entering the solution, and the concentration of sugar in the solution is so high, that the rate at which particles of sugar fall out of solution begins to increase. Eventually, if there is enough sugar, a balance point is reached and the solution is said to be saturated. At this point it seems that no more sugar is dissolving, but actually the rate at which it dissolves is now the same as the rate at which the solid reforms. The system is said to be at equilibrium. Like all equilibrium systems, this equilibrium can be affected by temperature.

An equilibrium can only be established in an enclosed system – one from which reactants and products cannot escape. Solids cannot escape from solutions, but gaseous reactions can only reach equilibrium in sealed containers.

CHEMICAL EQUILIBRIUM

Balance points are reached in many reversible chemical reactions, when neither the forward nor the backward reaction is complete.

If the reaction can go both ways, how do I know which is forward?

By convention the *reactants* are shown on the *left* side of the stoichiometric equation and the *products* on the *right*. The *left-to-right* reaction is the *forward* reaction.

The reaction between ethanol and ethanoic acid

Most people are familiar with ethanol as being the 'active' constituent of beers, wines and spirits, whereas ethanoic acid is usually associated with chips – it is in vinegar. When the two substances are mixed, they react slowly to form a compound, an ester called ethyl ethanoate: it has a distinctive sweet and fruity smell, and is used to make varnishes and glues. This **esterification** reaction, which is typical of alchohols and organic acids (see Module 7), is relatively easy to monitor and so has become a frequently used example in the study of chemical equilibrium.

- The forward reaction:

$$CH_3CH_2OH(l) + CH_3COOH(l) \rightarrow CH_3COOCH_2CH_3(l) + H_2O(l)$$
ethanol   ethanoic acid ethyl ethanoate   water

As the reaction proceeds so the chemicals are used up – the concentrations of the reactants decrease – and the reaction begins to slow down.

- The back reaction:

$$CH_3COOCH_2CH_3(l) + H_2O(l) \rightarrow CH_3CH_2OH(l) + CH_3COOH(l)$$

As the forward reaction slows down so the back reaction speeds up. As the concentrations of the products increase, they begin to react together to form ethanol and ethanoic acid.

Eventually, the rates of both reactions are equal – the system has reached a balance point and it is at **equilibrium**. Note though, the reaction *has not stopped* – it is just going equally as fast in both directions – the equilibrium is called a **dynamic equilibrium**.

All equilibrium processes, both physical and chemical, are dynamic.

It is important to note that when a system has reached equilibrium:

(a) Each substance is being produced by one reaction as fast as it is being used by the reverse reaction. So, at equilibrium the *concentrations* of the *reactants and products* do not change (see Figure 5.4.1).

$$CH_3CH_2OH + CH_3COOH \rightleftharpoons CH_3COOCH_2CH_3 + H_2O$$

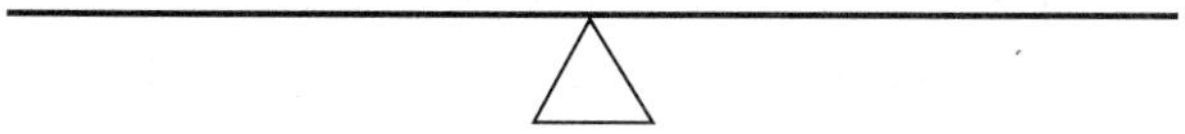

Figure 5.4.1
A system of equilibrium

(b) The forward and backward reactions still occur but at equal rates.

To show that a reaction can go in either direction the two equations can be put together and the single arrows $\rightarrow$ are replaced by the equilibrium symbol, a double arrow $\rightleftharpoons$, as in the esterification reaction:

$$CH_3CH_2OH(l) + CH_3COOH(l) \rightleftharpoons CH_3COOCH_2CH_3(l) + H_2O(l)$$

In practice, this reaction takes several weeks to reach equilibrium; the reaction can be speeded up by heating or by using a small amount of acid as a catalyst.

5.4.3 Factors Affecting Systems at Equilibrium

Equilibrium reactions are of theoretical interest to the chemist in the laboratory, who can provide valuable information about them to the chemical industry, where many are of vital importance. However, if some of the products continually recombine to reform starting substances, then product yields will be lower than anticipated and may make a commercial process uneconomical. Chemists have had to find ways of encouraging the forward reaction to proceed.

REMOVAL OF CHEMICALS CAN AFFECT THE REACTION

In the esterification reaction,

$$CH_3CH_2OH(l) + CH_3COOH(l) \rightleftharpoons CH_3COOCH_2CH_3(l) + H_2O(l)$$

it can be seen that if water was removed from the system then the equilibrium balance would be upset, as shown in Figure 5.4.2.

Figure 5.4.2
The removal of water shifts the position of equilibrium

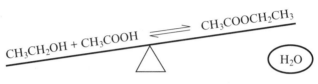

So if water is removed, say, by adding a powerful dehydrating agent, the ethyl ethanoate has nothing to react with and the reverse reaction is prevented. The equilibrium tries to re-establish itself by producing more ethyl ethanoate and water. If we continue removing water, the forward reaction can proceed until all of the reactants are converted to the ester.

The effects of changing conditions on a system at equilibrium can be predicted using **Le Chatelier's Principle**.

Le Chatelier's Principle states that:
If a system at equilibrium is disturbed by changing the conditions,
the system will react in such away as to oppose the change.

In other words, if something is done to the equilibrium system, the position of equilibrium will change in order to reduce or remove the effects of the disturbance, and to get back into balance.

EFFECTS OF PRESSURE CHANGES ON THE EQUILIBRIUM STATE IN GASEOUS REACTIONS

This phenomenon is well illustrated using the conversion of nitrogen and hydrogen to ammonia by the Haber Process:

$N_2(g) + 3H_2(g) \rightleftharpoons 2NH_3(g)$
1 mole   3 moles   2 moles

- The equation shows that a total of 4 moles of reacting gas produces 2 moles of ammonia.
- We know that the volume occupied by a gas at constant pressure is proportional to the number of moles of that gas (see Unit 4.5).

If the *pressure* on the reacting system is *increased*, Le Chatelier's Principle tells us that the system will try to *reduce* the pressure again by reducing the number of particles – so increasing the space available for them. The four moles of reacting gases occupy a larger volume than two moles of ammonia – so the increased pressure results in more ammonia being formed.

Increasing the pressure on this reaction also has the advantage of increasing the *concentration* of the gas molecules so the reaction will go faster.

THE EFFECT OF TEMPERATURE ON THE EQUILIBRIUM STATE

The effect of temperature on a system at equilibrium will depend on whether or not the reaction takes in or gives out heat during its course, *e.g.* in the reaction to produce ammonia:

$$N_2(g) + 3H_2(g) \rightleftharpoons 2NH_3(g) \quad \Delta H = -92 \text{ kJ mol}^{-1}$$

The reaction to produce ammonia is **exothermic** – it produces heat. Using Le Chatelier's Principle we can see that if this heat were removed, by using a lower temperature, it would 'encourage' the production of more ammonia (so releasing more heat to maintain the balance).

Unfortunately, although a low temperature increases the *yield* of ammonia, the reaction becomes uneconomically *slow*, as now there are fewer sufficiently energetic collisions between the particles. Obviously the industrial chemist faces the problem of reconciling both the kinetic and equilibrium considerations for this reaction to make economical to operate. In practice a compromise is reached by selecting an optimum operating temperature and pressure (see the Haber Process on page 248).

In general, when the temperature of a reacting system at equilibrium is raised:

- If the forward reaction is exothermic, increasing the temperature will tend to inhibit the reaction and decrease the concentration of products.
- If the forward reaction is endothermic, the system will tend to absorb the extra heat energy resulting in more products.

THE EFFECT OF CATALYSTS ON THE EQUILIBRIUM STATE

A catalyst can speed up both the forward and backward reactions and so enable the equilibrium state to be reached more quickly. Catalysts do not, however, affect the relative proportions (concentrations) of the reactants or products, so **they do not affect the position of equilibrium**.

5.4.4 Some Industrial Processes that involve Equilibrium

We will now look at some other important industrial applications of the chemical principles studied so far. Each of these processes involves at least one step that reaches equilibrium, and so the position of the equilibrium has to be altered in order to make the process economically viable. Although both raising the temperature and increasing the concentration of reactants can speed up reactions, the latter method is costly and the former may destroy some substances in the process. The use of a suitable catalyst is generally more efficient and economical in the long run. A considerable amount of time and money is invested in the development of catalysts. The detailed composition of an industrial catalyst is often a well-guarded secret.

THE HABER PROCESS

Ammonia is an important starting material in the production of ferti-lisers, nitric acid and polymers. The materials for making ammonia are air, natural gas (methane) and steam. Methane and steam are reacted over a nickel catalyst, at high temperature and 3 MPa (30 atm) pressure, to give carbon monoxide and hydrogen. Air is then added to achieve the correct ratio of hydrogen to nitrogen. The carbon monoxide is removed by a catalytic reaction with more steam to make it into carbon dioxide, which can then be removed by dissolving it in a variety of chemicals.

The production of ammonia by the Haber Process utilises all four factors that affect equilibrium reactions:

a) We have seen that raising the temperature in this reaction inhibits production of ammonia, but the reaction is slow at low temperature. An optimum yield is attained by using a compromise temperature of about 500 °C.
b) High pressure improves yield, but the cost of building and running a plant that can withstand very high pressure is very expensive. In practice, both the rate of production and yields are improved at a moderate operational pressure of 15–30 MPa (150–300 atm).
c) Iron, with some potassium oxide or aluminium oxide added as a promoter, is used as the catalyst (see page 235).
d) Ammonia is removed as it forms – when cooled it condenses easily at high pressure – which encourages a higher yield of product.

Only 15–20% of the gases react each time that they pass through the reactor; unreacted hydrogen and nitrogen are recycled, with more being fed in to maintain the pressure. Once started, the process is continuous until the whole plant is closed for maintenance.

THE CONTACT PROCESS

Sulfuric acid is used in many important manufacturing processes. Some examples are:

- pharmaceuticals/insecticides;
- making detergents;
- cleaning steel before galvanising or 'tinning', *etc.*
- electrolyte in car batteries;
- paint manufacture;
- making fertilisers;
- extracting metals from their ores;
- manufacture of rayon and other fibres;
- making paper.

In 1990, the total world-wide output of sulfuric acid, *per day,* was about 200 000 tonnes.

Production begins with the burning of molten sulfur in dry air to produce sulfur dioxide:

$$S(l) + O_2(g) \rightarrow SO_2(g)$$

An initial input of heat is needed to get the reaction started, but then it is so exothermic that the reaction is self-sustaining, and some of the excess energy released can be used to produce steam at high pressure that can in turn power pumps and provide heat elsewhere in the plant.

The second and key part of this process is also the most problematic:

$$2SO_2(g) + O_2(g) \rightleftharpoons 2SO_3(g) \quad \Delta H = -197 \text{ kJ mol}^{-1}$$

The conversion of sulfur dioxide to sulfur trioxide is slow and involves a chemical equilibrium. If we apply Le Chatelier's Principle, we can see that the forward process would be favoured by:

a) a low temperature;
b) high pressure.

However, in commercial processes scientific theory must be tempered with economics. Cooling the reacting system would make the reaction unprofitably slow, and building chemical plants that operate at high pressures is very costly. The problems of economic production are overcome by:

a) **Using a catalyst**. Vanadium(V) oxide (V_2O_5) is the principal catalyst; although less efficient than the platinum catalysts used in earlier years, it is cheaper and less easily poisoned by impurities in the sulfur dioxide. The catalyst is said to be surface active; it 'holds' the molecules of sulfur dioxide and oxygen at its surface long enough for the reaction to take place (adsorption). The product which is released or 'desorbed', is sulfur trioxide (Figure 5.4.3). Vanadium(V) oxide is mixed with other substances and formed into cylinders which offer a large surface and allow good flow of gases.

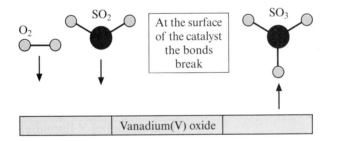

Figure 5.4.3
V_2O_5 is used as a heterogeneous catalyst in the Contact Process

b) **A compromise temperature** of 450 °C is used; at this temperature the forward reaction is fast and the back reaction slow: 97% of the sulfur dioxide can be converted to products. A lower temperature also has the dual benefits of keeping production costs down and minimising corrosion in the reaction vessels.
c) **A pressure of a few atmospheres** is needed to keep the gases moving through the reactors.
   Sulfur trioxide is not dissolved directly into water as the reaction is violent and would produce corrosive gases.

$$SO_3(g) + H_2O(l) \rightarrow H_2SO_4(aq)$$

Instead, the final stage involves the gas being slowly absorbed into concentrated sulfuric acid, to produce 'oleum' or 'fuming sulfuric acid' – an oily liquid – ($H_2S_2O_7$) which can be mixed with dilute sulfuric acid to produce the final product of the required concentration:

$$SO_3(g) + H_2SO_4(l) \rightarrow H_2S_2O_7(l)$$

$$H_2S_2O_7(l) + H_2O(l) \rightarrow 2H_2SO_4(l)$$

As we have seen on pave 247, chemical equilibria are affected by temperature so the value for the equilibrium constant, K_c, is quoted for a specific temperature, often 25 °C (298 K).

By convention the concentrations of the *products* are always the *numerator* (*i.e.* the *upper* line of the equation). So it is important to relate the equilibrium constant to the appropriate equation for the reaction.

When equilibrium constants are known they provide information about the probable outcome of a reaction (the extent to which products have been formed), but care must be taken when using them in discussion.

It is important to remember that:

a) The value of the equilibrium constant is determined by the extent of the reaction at equilibrium; *i.e.* the proportion of products and reactants in the equilibrium mixture: it is *not* an indicator of the *rate of reaction*.

b) The Equilibrium Law only applies to systems at equilibrium.

c) K_c is constant for *a given temperature only*.

d) Catalysts speed up the rate of chemical reactions, **but they have no effect on the position of equilibrium** nor, therefore, on the value of the equilibrium constant K_c.

The order of magnitude of equilibrium constants varies over a huge range, depending on the particular reaction:

• A high value of K_c such as 3.4×10^{17} indicates mostly products in the equilibrium mixture.

• $K_c = 1$ indicates equal concentrations of reactants and products in the equilibrium mixture.

• A very low value of K_c such as 3×10^{-18} indicates that at equilibrium very few of the reactants have been converted to products.

NOTE: No units have been given for the value of the equilibrium constant, K_c, and this is often the case. The equilibrium constant *will* have units however, if, in the equation:

$$K_c = \frac{[C]^c\,[D]^d}{[A]^a\,[B]^b}, \quad (c + d) \text{ does not equal } (a + b).$$

This occurs because concentrations are measured in mol dm^{-3}, so, for example, in the Haber Process, K_c will be:

$$N_2(g) + 3H_2(g) \rightleftharpoons 2NH_3(g)$$

$$K_c = \frac{[NH_3]^2}{[N_2]\,[H_2]^3}$$

Looking only at the units involved:

$$K_c = \frac{[\text{mol dm}^{-3}]^2}{[\text{mol dm}^{-3}]\,[\text{mol dm}^{-3}]^3}$$

$$K_c = \frac{1}{[\text{mol dm}^{-3}]^2} = \text{mol}^{-2}\,\text{dm}^6.$$

Answers to diagnostic test

If you score less than 80%, then work through the text and re-test yourself at the end by using this same test. If you still have a low score then re-work the topic at a later date.

1 If an equilibrium is disturbed by changing the conditions, the position of the equilibrium moves to try to counteract that change. **(1)**

2 a) The equilibrium would <u>shift to the right</u>, *i.e.* <u>more silver would be deposited</u> and the solution would look 'brownish green'. **(2)**

b) The equilibrium would <u>shift to the left</u>, *i.e.* <u>some silver would dissolve</u>. **(2)**

3 a) Thermal dissociation is a <u>reversible reaction</u> which is dependent on temperature, *i.e.* heating or cooling can reverse the process such as in the dissociation of ammonium chloride into ammonia and hydrogen chloride: **(1)**

$$NH_4Cl(s) \rightleftharpoons NH_3(g) + HCl(g)$$

Other common examples are given below: award yourself a mark for any one. **(1)**

$$2HgO(s) \rightleftharpoons 2Hg(l) + O_2(g)$$

$$CaCO_3(s) \rightleftharpoons CaO(s) + CO_2(g)$$

$$N_2O_4(g) \rightleftharpoons 2NO_2(g)$$
Dinitrogen   Nitrogen
tetroxide   dioxide

b) Thermal decomposition occurs when substances are completely '<u>broken down</u>' by heat, a process which <u>cannot be easily reversed</u>. **(1)**

Four common examples are given: award a mark for each correct formula and number of moles.

$$2KClO_3(s) \rightarrow 2KCl(s) + 3O_2(g)$$
Potassium   Potassium
chlorate   chloride

$$NH_4NO_3(s) \rightarrow N_2O(g) + 2H_2O(g)$$
Ammonium   Dinitrogen
nitrate   oxide

$$2Pb(NO_3)_2(s) \rightarrow 2PbO(s) + O_2(g) + 4NO_2(g)$$
Lead(II)   Lead(II)
nitrate   oxide

$$2AgNO_3(s) \rightarrow 2Ag(s) + O_2(g) + 2NO_2(g)$$
Silver   Nitrogen
nitrate   dioxide **(1)**

4 a) An <u>exothermic reaction generates heat</u>. The equilibrium position <u>shifts to the left</u> to resist the increase in temperature, and so <u>less ammonia</u> is produced.   **(3)**

   b) There are only <u>2 moles of products</u> on the right-hand side of the equation, which will then <u>occupy a smaller volume</u>. The equilibrium position shifts to the right to resist the increase in pressure: <u>more ammonia is produced</u>.   **(3)**

5 a) The equation for the reaction between ethanol and ethanoic acid is:

$$C_2H_5OH(l) + CH_3COOH(l) \rightleftharpoons CH_3COOC_2H_5(l) + H_2O(l)$$ **(2)**

   (Left side of equation correct = 1 mark; right side of equation correct = 1 mark)

   b) The equilibrium constant is given by

$$K_c = \frac{[Products]}{[Reactants]} = \frac{[CH_3COOC_2H_5] \times [H_2O]}{[C_2H_5OH] \times [CH_3COOH]}$$ **(3)**

   (1 mark each for correct symbol for equilibrium constant; numerator and denominator)

20 Marks   (80% = 16)

Unit 5.5
Quantitative Aspects of Acid–Base Equilibria

Starter check

What you need to know before you start this unit.

You will need a good understanding of the previous units of Module 5 and also Module 4.

Aims **By the end of this unit you should understand and be able to:**

- Distinguish between an acid or alkali by its pH.
- Do simple pH calculations.
- Explain the effects of acids and alkalis on buffer solutions.

Diagnostic test

Try this test at the start of the unit. If you score more than 80%, then use this unit as a revision for yourself and scan through the text. If you score less than 80%, then work through the text and re-test yourself at the end by using this same test.

The answers are at the end of the unit.

1 Match the statements to the appropriate answer　　　　　**(6)**

Strong acids	have several moles of acid per dm^3 of solution
Dilute acids	are partially dissociated into ions in dilute solution
Weak acids	around pH = 3
The pH of 0.1 M hydrochloric acid is	are completely dissociated into ions in dilute solution
Concentrated acids	have few moles of acid per dm^3 of solution
The pH of 0.1 M ethanoic acid is	pH = 1

2 What does the pH of an acid actually measure?　　　　　**(2)**

3 a) What is the pH of a solution containing 10^{-13} mol dm^{-3} hydrogen ions? Is the solution acidic or alkaline? What colour would universal indicator be in this solution?　　　**(3)**

b) What is the pH of a solution containing 0.0001 mol dm^{-3} of hydrogen ions? Is the solution acidic or alkaline? What colour would you expect litmus indicator to be in this solution?   (3)

4 What is a buffer solution?   (1)

5 Complete the table   (5)

Type of buffer	Use	Made from
Acidic buffer		Weak acid plus the _____ of weak acid
	Maintains pH > 7	

20 Marks   (80% = 16)

5.5.1 Introduction

The next section builds on and uses the understanding of chemical equilibrium acquired in Unit 5.4. It looks again at the properties of acids and alkalis, and applies the principles of chemical equilibrium to the regulation of hydrogen ion concentration in living systems.

5.5.2 The Dissociation of Water

Even very pure distilled water will still *just* conduct electricity, so *water must contain ions*: it is said to *dissociate* into ions. The conductivity of pure water is very low, so there cannot be many ions in a given volume: the dissociation is said to be *partial* or incomplete.

$$H_2O(l) \rightleftharpoons H^+(aq) + OH^-(aq)$$

The equilibrium constant, K_c, for this reaction can be written:

$$K_c = \frac{[H^+(aq)] \times [OH^-(aq)]}{[H_2O(l)]} \text{ mol dm}^{-3}$$

It sounds like nonsense to talk about the concentration of water; indeed, there are so few ions present in a given volume, that the concentration of the liquid water hardly changes. To all intents and purposes the concentration of water molecules present remains constant; so this constant is incorporated with K_c, and we usually write the equilibrium constant like this,

$$K_c \times [H_2O] = [H^+(aq)] \times [OH^-(aq)]$$

so: $K_c \times$ another constant $= [H^+(aq)] \times [OH^-(aq)]$

or $K_w = [H^+] \times [OH^-]$

(where K_w is the equilibrium constant for water, called the **ionic product of water**).

Experimentation shows that at room temperature, 1 dm^3 of pure water contains only 0.000 0001 or 1×10^{-7} moles of hydrogen ions. So the *concentration* of hydrogen ions in pure water is 1×10^{-7} mol dm^{-3}.

Pure water is neutral, so the number of H$^+$ ions and OH$^-$ ions must be equal.

At 25 °C (298 K) $[H^+] = [OH^-] = 1 \times 10^{-7}$ (mol dm^{-3})

hence $\qquad\qquad K_w = [H^+] \times [OH^-] = 1 \times 10^{-7} \times 1 \times 10^{-7}$
$\qquad\qquad\qquad\qquad = 1 \times 10^{-14} \text{ mol}^2 \text{ dm}^{-6}$

This is a very small value for an equilibrium constant, as we might expect since pure water is only very partially ionised. In fact, at 25 °C only one water molecule in about 550 million is ionised at any given moment! Yet this is *vitally* important – life would not be possible otherwise.

5.5.3 The Measurement of pH, and the pH Scale

CALCULATING VALUES OF pH

The strength of an acid is indicated by its pH: this is a sort of 'upside-down' measure of the concentration of hydrogen ions in solution. As the pH goes up, so the concentration of hydrogen ions in solution, $[H^+]$, goes down. The 'p' part of pH comes from the German word 'potenz' meaning power.

In pure water and in neutral aqueous solutions such as sodium chloride, any hydrogen or hydroxide ions present come *only* from the ionisation of water. Hence:

$[H^+] = [OH^-] = 1 \times 10^{-7}$ mol dm^{-3}
for convenience this is often written as 10^{-7}.

Let us stop for a moment and consider the size of the numbers we are using. Every time that we want to state the strength of an acid we need, at best, to express the concentration of H^+ ions in standard index form or write out a string of noughts. If, however, we use the base 10 logarithms (logs) of very large or very small numbers they become much more manageable (see Module 4).

By definition, $\log_{10} 0.1 = \log_{10} 10^{-1} = -1$

and $\log_{10} 0.001 = \log_{10} 10^{-3} = -3$, *etc.*

And don't forget! $\log_{10} 1 = \log_{10} 10^0 = 0$

- pH is defined as the negative logarithm to base 10 (written as log) of the hydrogen ion concentration, $[H^+]$, in a solution. This convention means that chemists can use small positive numbers to express $[H^+]$ instead of working in powers of ten.

pH = −log [H⁺]
so the pH of pure water = −log [10⁻⁷] which is 7
so *any* solution with pH 7 is neutral.

- On your calculator input **0.000 000 1**, then press **[log] [=]**

  What is the result?

- You can also use the exponential EXP button on the calculator:

  [1] [EXP] [+/−] [7] [log] [=]

e.g.
- In 1 dm³ of 0.1 mol dm^{-3} nitric acid (or 0.1 M − see Module 4) there are 0.1 or 10^{-1} moles of hydrogen ions.
- Nitric acid of concentration 0.1 mol dm^{-1} therefore has a pH of 1 − it has a low pH.
- Dilute nitric acid is *completely ionised*,

$HNO_3(aq) \rightarrow H^+(aq) + NO_3^-(aq)$

- In 1 dm³ of 0.1 mol dm⁻³ ethanoic acid, however, there are only 0.001 or 10^{-3} moles of hydrogen ions.
- Ethanoic acid of concentration 0.1 mol dm⁻³ has a pH of 3.
- This is because ethanoic acid is only *partially ionised*:

$$CH_3COOH(aq) \rightleftharpoons CH_3COO^-(aq) + H^+(aq)$$

CALCULATING THE pH OF ALKALINE SOLUTIONS

It is easy to see how the pH of an acid is worked out, because by definition an acid is a proton (H^+) donor; but what about alkalis?

A solution of sodium hydroxide contains 0.1 moles of NaOH per dm³.

- It has a *concentration* of 0.1 mol dm⁻³.

- Sodium hydroxide is a strong alkali – it is completely ionised.

- So $[OH^-]$ ions $= 10^{-1}$ mol dm⁻³ (remember that [X] means concentration of X, in mol dm⁻³).

- But, for an aqueous solution, the ionic product of water is constant.

- $[H^+] \times [OH^-] = 10^{-14}$ mol² dm⁻⁶

- so $[H^+] = \dfrac{10^{-14}}{[OH^-]} = \dfrac{10^{-14}}{10^{-1}} = 10^{-13}$ mol dm⁻³

$[H^+] = 10^{-13}$ indicates that there are very few hydrogen ions in the solution

- $pH = -\log [H^+] = -\log 10^{-13} = 13$

∴ Sodium hydroxide solution of concentration 0.1 mol dm⁻³ has a pH of 13.

Generally:

Neutral solutions $[H^+] = [OH^-] = 10^{-7}$ mol dm⁻³ and pH = 7 at 25 °C (298 K)
Acidic solutions $[H^+] > [OH^-]$, and pH < 7 at 25 °C
Alkaline solutions $[H^+] < [OH^-]$, and pH > 7 at 25 °C

Try these calculations

1 What is the pH of a solution containing 0.0001 (10^{-4}) moles H^+ ions?
2 What is the pH of a solution of hydrochloric acid [HCl(aq)] containing 0.02 moles H^+ ions per dm³?
3 What is the pH of a solution containing 5×10^{-4} mol dm⁻³ H^+ ions?
4 What will be the pH of 1.7×10^{-5} mol dm⁻³ HCl(aq)?
5 A solution of a strong base contains 0.01 mol dm⁻³ hydroxide ions (OH^-). What is the pH of the solution?

Answers

1 pH = 4
2 pH = 1.698 = 1.7
3 pH = 3.3
4 pH = 4.769 = 4.8
5 pH = 12

5.5.4 Detecting Acidity and Alkalinity

Acids are substances that dissociate in water to produce hydrogen ions. This idea was first put forward by the Swedish chemist Arrhenius in the nineteenth century. For example, although hydrogen chloride gas has a covalent structure, it dissolves in water to form:

$$HCl(g) + water \rightarrow H^+ (aq) + Cl^-(aq)$$

As knowledge about atomic structure developed, chemists realised that the hydrogen ion was simply a proton, and it was highly unlikely to be stable on its own. There is now considerable evidence, from electrolysis reactions, to suggest that hydrogen ions combine with polar water molecules to form stable oxonium ions:

$$H^+(aq) + H_2O(l) \rightarrow H_3O^+(aq)$$

When hydrogen chloride dissolves in water, there is thought to be a chemical reaction:

$$HCl(aq) + H_2O(l) \rightarrow H_3O^+(aq) + Cl^-(aq)$$

The same holds true for the other common acids: they all dissociate, to a greater or lesser extent, in water to form oxonium ions. This idea helps to explain why *acids only show their acidic nature when dissolved in water*.

Traditionally, acids have been defined as substances that:

- produce hydrogen gas on reaction with metals;
- neutralise bases to give salt and water as the only products;
- release carbon dioxide when reacted with carbonates.

This information conveniently summarises the properties of most acids, but does not tell us anything about their structures and differences in reactivity. The evidence for the ionisation of acids (such as conductivity experiments) gives us an important extra property to include in a full definition – *acids provide hydrogen ions, H⁺ (aq), in solution*. **Acids are proton donors**, and since *alkalis* are the chemical opposites of acids, they must be *proton acceptors*.

Here we should remind you about a common notation you will often see. It has just been said that the H^+ ion is thought to combine with water to form the more stable hydroxonium ion,

$$H^+(aq) + H_2O(l) \rightleftharpoons H_3O^+(aq)$$

so the symbols H^+, $H^+(aq)$ and H_3O^+ all refer to hydrogen ions in solution, and one or the other or even all may be used by different text books. For simplicity, this book will continue to use $H^+(aq)$.

THE FORMATION OF HYDROGEN IONS, AND THE ACID DISSOCIATION CONSTANT

Experiment shows us that some acids dissociate to a greater extent than others, *i.e.* more of their molecules produce H^+ ions and the equilibrium position is further to the right. For a general acid of formula HA:

$$HA(aq) + H_2O(l) \rightleftharpoons H^+(aq) + A^-(aq)$$

$$\text{or } H_3O^+(aq)$$

By Le Chatelier's Principle (Section 5.4.3), adding more water, *i.e.* diluting the acid, will *increase the ionisation* of the acid.

The acid dissociation constant should be:

$$K_a = \frac{[H^+][A^-]}{[HA][H_2O]}$$

but the concentration of the water present will hardly vary, so we can incorporate that constant into K_c, to give –

$$K_a = \frac{[H^+][A^-]}{[HA]} \text{ mol dm}^{-3}$$

where K_a is the acid dissociation constant.

- The greater the degree of ionisation, the stronger the acid – the higher the value of K_a.
- For acids of the same concentration, the greater the degree of ionisation of the acid, the more ions are present, and so the higher the conductivity.

Some typical values of acid dissociation constants are given in Table 5.5.1:

Table 5.5.1 *Acid dissociation constants for some common organic acids*

Acid	Formula	Acid dissociation constant, K_a (mol dm^{-3})
Ethanoic	CH_3COOH	1.8×10^{-5}
Trichlorethanoic	CCl_3OOH	2.3×10^{-1}
Benzoic	C_6H_5COOH	6.3×10^{-5}
Methanoic	$HCOOH$	1.6×10^{-4}

Why is the value of K_a for trichloroethanoic acid more than 10 000 *times larger than that for ethanoic acid?*

Chlorine is an electronegative atom; it pulls electrons towards its nucleus. This effect is felt throughout the molecule, and thus the O–H bond is weakened with the result that hydrogen ions are more easily formed. More hydrogen ions in solution means a stronger acid.

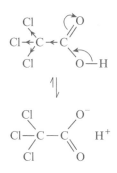

The values of K_a for the common inorganic acids, such as sulfuric and nitric acids, are of course very high because the acids are fully ionised in dilute solution. They are called *strong acids*.

You will also come across references to pK_a (*cf.* pH, Section 5.5.3):
$$\text{where } \mathbf{pK_a = -\log K_a}$$
Using this definition – **the smaller the value of pK_a, the stronger the acid.**

STRONG AND WEAK ACIDS

- **Hydrochloric acid** is said to be a **strong acid** as it is completely dissociated in dilute solution:

$$HCl(aq) \rightarrow H^+(aq) + Cl^-(aq)$$

- **Ethanoic acid** and **phenol** are **weak acids** as they are never completely ionised, however dilute the solution**.**

  Phenol is a *weak* acid: $C_6H_5OH(aq) \rightleftharpoons C_6H_5O^-(aq) + H^+(aq)$

  Ethanoic is a *weak* acid: $CH_3COOH(aq) \rightleftharpoons CH_3COO^-(aq) + H^+(aq)$

Many weak acids occur naturally and are organic in origin. For instance:

- methanoic acid (which used to be called formic acid) is present in the sting of ants;
- ethanoic acid (formerly called acetic acid) is in vinegar – it has a characteristic and very pungent smell;
- butanoic acid gives rancid butter and cheese its characteristic odour.

N.B. Do not confuse the terms **STRONG** and **CONCENTRATED**.

- strong acids are **completely dissociated** into ions in dilute solutions;
- concentrated acids (or any concentrated solution) contain **several moles of substance per dm³** of solution.
- ordinary 'lab' concentrated hydrochloric acid contains approximately 10 mol dm^{-3} of HCl(aq) and concentrated sulfuric acid contains about 18 mol dm^{-3} $H_2SO_4(aq)$.

The concept of 'strong' and 'weak' may be applied to any electrolyte to indicate the degree to which they dissociate (break up) into ions.

THE USE OF THE pH SCALE

The acidity, or alkalinity of a solution, is frequently assessed using universal indicator paper or solution, which is a mixture of a variety of indicators that go through a spectrum of colours as the acidity of a solution changes. In Table 5.5.2, pH values have been matched to H^+ ion concentration and the usual colour of universal indicator.

Table 5.5.2 *Relation between pH, [H⁺] and the colour of universal indicator*

pH scale	0	1	2	3	4	5	6	7	8	9	10	11	12	13	14
[H⁺]	1	10^{-1}	10^{-2}	10^{-3}	10^{-4}	10^{-5}		10^{-7}			10^{-10}	10^{-11}			10^{-14}
Colour of universal indicator	Red			Orange		Yellow			Green		Blue			Purple	
			Acid						Neutral					Alkaline	

5.5.5 The Effect of Strong Acids and Alkalis on Living Organisms

All strong acids can be dangerous; high concentrations of H^+ ions will, for example, break down large protein molecules in the skin and destroy chemicals in the body. For instance, concentrated nitric can often oxidise organic material – *yes, and that means you* – to carbon dioxide. Concentrated sulfuric acid – which is a strong dehydrating agent as well – can convert sugar into a pile of carbon in seconds and, given half a chance, will do the same to your skin or eyes. The acid in your stomach is naturally very strong; it has a pH value between 1 and 2. Ulceration can occur if the stomach produces too much acid to cope with, say, after over-eating, and many people need to take antacids/indigestion remedies to neutralise partially the effects of the acid. These products often contain substances like sodium hydrogencarbonate (bicarbonate of soda). This reacts with excess acid in the stomach and re-establishes the correct acid balance.

BUFFER SOLUTIONS

A buffer solution keeps its pH almost constant by resisting changes in pH when small amounts of acid or alkali are added to it.

Buffer solutions play an important role in many processes where the pH of a system needs to be maintained at an optimum value. For instance, several synthetic and processed foods contain buffers so that they may be digested without causing undue changes in the chemistry of the body. Buffer solutions are also important in agriculture and medicine; for example, intravenous injections are carefully 'buffered' so as not to alter the blood pH from its normal value of 7.4.

The main use of buffers in the laboratory is in the preparation of solutions of *known and constant* pH. It is difficult to ensure that the pH of a solution is accurate simply by preparing an acid or alkali of a given concentration because, for instance, atmospheric gases such as carbon dioxide dissolve in them, and so the pH will vary slightly over a period of time.

Water is a major constituent of all living organisms, and the seas support millions of different species of plant and animal life. Yet the pH values of both water and saline (salt) solutions are particularly sensitive to the addition of acids or alkalis. Look:

- If just 0.1 cm³ of 1.0 mol dm^{-3} hydrochloric acid is added to 1 dm³ of water the pH falls sharply from 7.0 to 4.0, *i.e.* by 3 pH 'points'.

Why?

- 1 dm³ of 1 mol dm^{-3} acid contains 1 mole H$^+$ ions.
- ∴ 0.1 cm³ of 1 mol dm^{-3} acid contains 1/10 000 mol H$^+$ ions.
- So the water now contains 1/10 000 (10^{-4}) mol H$^+$ ions.
- Remember, pH = $-\log_{10}$ [H$^+$]
- and $-\log_{10} 10^{-4} = 4.0$
- ∴ pH of new solution = 4.

Obviously, there must be systems in place that prevent the delicate pH balance of living systems from being fatally upset, such as in Figure 5.5.1. Every system is different, but the general principles for the action of buffers are the same.

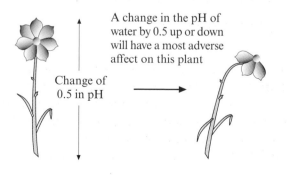

A change in the pH of water by 0.5 up or down will have a most adverse affect on this plant

Change of 0.5 in pH

Figure 5.5.1
Plants are very sensitive to changes in pH

HOW DO BUFFER SOLUTIONS WORK?

There are two different types of buffer solutions:

- Acidic buffers work to keep the pH of the system below 7.0.
- Alkaline buffers work to keep the pH of the system above 7.0.

Clearly each buffer will work in a specific way to maintain the pH of a system, but it is easy to predict what will happen by applying Le Chatelier's Principle.

Acidic buffer solutions

These consist of a weak acid and a salt of that acid.

e.g. Ethanoic acid and sodium ethanoate.
    Ethanoic acid is only *partially dissociated* into ions:

$$CH_3COOH(aq) \rightleftharpoons CH_3COO^-(aq) + H^+(aq)   \textbf{(1)}$$

Sodium ethanoate is an ionic compound; it is *completely dissociated* into ions:

$$CH_3COONa(aq) \rightarrow CH_3COO^-(aq) + Na^+(aq)   \textbf{(2)}$$

- When a small amount of a strong acid is added is added to the solution containing the buffer, there is an excess of H^+ ions which upsets the position of the equilibrium in **(1)** and, in accordance with Le Chatelier's Principle, causes a shift to the left in order to remove those ions. A weak acid such as ethanoic can only remove so many H^+ ions by shifting its equilibrium position: the addition of an ionic salt to a buffer solution, in this case sodium ethanoate **(2)**, produces many more ethanoate ions which can then combine with any excess H^+ ions [not removed by **(1)**]. This forms the weak ethanoic acid, and so restores the pH of the solution.
- When an **alkali** is added to the solution containing the buffer, there is an excess of OH^- ions which tend to react with the free H^+ ions, forming neutral water and so they are removed from the system. The equilibrium in **(1)** then moves to the right, which has the effect of producing more H^+ ions to 'mop up' the hydroxide ions.

Alkaline buffer solutions

In the laboratory and in industry there is a need for buffer solutions to maintain the alkalinity of systems. An **alkaline buffer** solution is a mixture of a weak alkali and its salt, such as a solution of aqueous ammonia with ammonium chloride. Ammonia in water is partially protonated **(3)** and its salt is fully dissociated **(4)**. Alkaline buffer solutions work in a similar way to acidic ones to keep the pH of a system *above* 7.

The addition of small amounts of acid (H^+ ions) to the buffer causes a shift in the equilibrium position of reaction **(3)**, effectively 'mopping up' the acid and restoring the pH of the system.

Similarly, any excess of hydroxide ions is removed because their presence causes the equilibrium position of reaction **(3)** to shift to the left, removing OH^- ions from the system. Any remaining excess of OH^- ions reacts with the large quantities of ammonium ions, restoring the pH of the system.

$NH_3(aq) + H_2O(l) \rightleftharpoons NH_4^+(aq) + OH^-(aq)$ **(3)**

$NH_4Cl(aq) \rightarrow NH_4^+(aq) + Cl^-(aq)$ **(4)**

It can be seen that the components in a buffer mixture work both individually and to complement each other. There are, though, *limits* to the effectiveness of a buffer solution; if either alkali or acid is added in excess, it will destroy the action of the buffer. We say that the buffer has a *buffer capacity*, which is equivalent to the amount of acid or alkali that can be added before the pH of the buffer changes.

5.5.6 Calculating the pH of a Buffer Solution

An acidic buffer solution is made up of a weak acid HA and its salt (M^+A^-), where 'H' denotes the presence in the molecule of hydrogen atoms which can form ions, and 'M^+' refers to a (usually) metal ion:

$HA(aq) \rightleftharpoons H^+(aq) + A^-(aq)$ *partially ionised*

$MA(aq) \rightarrow M^+(aq) + A^-(aq)$ *fully ionised*

The expression for the acid dissociation, K_a, constant can be written:

$$K_a = \frac{[H^+][A^-]}{[HA]}$$

so $[H^+] = K_a \times \dfrac{[HA]}{[A^-]}$

In a buffer mix the value of [HA] is effectively the initial concentration, of the acid since weak acids are only very slightly dissociated, and [A$^-$] is effectively the initial concentration of the salt because it is completely ionised.

Thus we can write:

$$[H^+] = K_a \times \frac{[\text{acid}]}{[\text{salt}]}$$

So pH $= -\log_{10}[H^+]$

$$= -\log_{10}(K_a \times \frac{[\text{acid}]}{[\text{salt}]})$$

For an extension of this topic, see Section 5.5.8 on page 270.

5.5.7 The Effects of Buffers in the Body

Hydrogen ions are constantly being made by the body, but blood and tissue fluids are held at about pH 7.4. This balance is achieved in three ways:

a) During respiration, the lungs expel carbon dioxide which would otherwise accumulate and combine with water to form carbonic acid.
b) The kidneys selectively excrete hydrogen ions and retain hydrogen carbonate (HCO_3^-) ions.
c) Buffering mechanisms in the blood suppress an increase in hydrogen ion concentration.

This section briefly looks at the chemistry of biological buffer systems.

Normal bodily functions such as digestion and exercise tend to produce H^+ ions rather than OH^- ions, so the most important function of buffers in the body is to regulate hydrogen ion concentration. To maintain the pH of the blood around 7.4 the body has a number of efficient buffer systems which usually consist of a weak acid and a weak base.

THE HYDROGENCARBONATE SYSTEM

This system is unusual in that it consists of *only* the salt sodium hydrogencarbonate, a *weak* base, which is also is the salt of the *weak* carbonic acid. In solution, sodium hydrogencarbonate readily dissociates into ions:

$$NaHCO_3(aq) \rightarrow Na^+(aq) + HCO_3^-(aq)$$

If there is an excess of free hydrogen ions in the body, they are immediately taken up by the hydrogencarbonate ions, to form *weak* carbonic acid which has no ill effects.

- **Add hydrogen ions** $H^+(aq) + HCO_3^-(aq) \rightarrow H_2CO_3(aq)$

If the alkalinity of the system is increased, hydrogencarbonate ions can react with free hydroxide ions to form carbonate ions and water.

- **Add hydroxide ions** $OH^-(aq) + H_2CO_3(aq) \rightarrow H_2O(l) + HCO_3^-(aq)$

This system helps to ensure that the pH of body fluids is never drastically altered. However, although in the body sodium hydrogencarbonate acts as a buffer on its own, it plays a relatively minor role as a buffer, and in most cases two or more substances are involved, as in inorganic systems. Phosphate salts play a more important role in regulating hydrogen ion concentration in the blood.

THE PHOSPHATE BUFFER SYSTEM

In this system the weak acid is sodium dihydrogenphosphate (NaH_2PO_4) and the weak base is disodium hydrogenphosphate (Na_2HPO_4). The action of the buffer system is similar to that of the hydrogencarbonate system.

- **Add acid** $\quad H^+(aq) + HPO_4^{2-}(aq) \rightarrow H_2PO_4^-(aq)$

- **Add alkali** $\quad OH^-(aq) + H_2PO_4^-(aq) \rightarrow HPO_4^{2-}(aq) + H_2O(l)$

These two buffer systems are normally sufficient to maintain the blood pH between 7.35 and 7.45; however, there are certain organic compounds, such as amino acids and haemoglobin, that are still more effective as they are in higher concentrations in the body.

THE AMINO ACIDS AS BUFFERS (see also Section 7.3.6)

The formula of glycine, NH_2CH_2COOH, gives no indication as to why it is a white crystalline solid at room temperature (compare this with propanoic acid, CH_2CH_2COOH, which is a liquid). The structure is explained once we know that in the crystal form glycine is ionic, in fact a dipolar ion – it carries both negative and positive charges (Figure 5.5.2). It is the attraction between these opposite charges on neighbouring molecules which makes glycine a solid.

Many other amino acids exist as dipolar ions or *zwitterions* (which means 'two ions'). These dipolar structures exist even when the acids dissolve in water, and are responsible for some unusual properties.

Amino acids are the molecular units from which large macromolecules, known as proteins, are made.

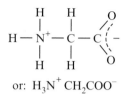

or: $H_3N^+ CH_2COO^-$

Figure 5.5.2
The structure of glycine

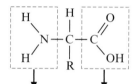

Amino group   Acid group

Figure 5.5.3
The general structure of an amino acid

The general structure of an amino acid

The general structure of an amino acid is shown in Figure 5.5.3.

Two amino acid molecules can join by a condensation reaction (Figure 5.5.4) to form a peptide link – the beginnings of a protein molecule (see Section 7.3.4). A protein may consist of 100 or more amino acid units, and may contain up to 20 different amino acids.

Proteins have both acidic and basic properties because of their amino (basic) and carboxyl (acidic) groups at the ends of the peptide chain. It is therefore possible for proteins to combine with both acids and bases, enabling them to function as buffers.

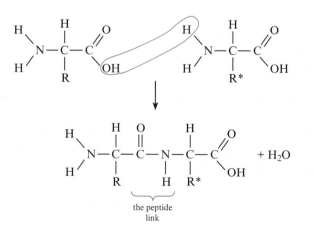

Figure 5.5.4
Joining amino acid units by a condensation reaction

Like ordinary amines, amino acids are *proton acceptors*, but it is *not* the lone pair on the nitrogen atom that holds the H^+ ion, because in free amino acids at neutral pH the nitrogen atom is already protonated.

When an amino acid comes into contact with a *strong acid*, such as hydrochloric acid, it is the ionised carboxyl group (COO^-) that accepts the proton and helps to regulate the acid balance.

- **Add acid** $^+NH_3CH_2COO^- + H_3O^+ \rightleftharpoons {}^+NH_3CH_2COOH + H_2O$

If a strong alkali such as sodium hydroxide is added to an amino acid solution, the protonated amino group becomes the acid (a proton donor) and releases a proton to mop up hydroxide ions.

- **Add alkali** $^+NH_3CH_2COO^- + OH^- \rightleftharpoons NH_2CH_2COO^- + H_2O$

Answers to diagnostic test

If you score less than 80%, then work through the text and re-test yourself at the end by using this same test. If you still have a low score then re-work the topic at a later date.

1 Strong acids	*are completely dissociated in to ions in dilute solution*	
Dilute acids	*have few moles of acid per dm³ of solution*	
Weak acids	*are partially dissociated into ions in dilute solution*	
The pH of 0.1 M hydrochloric acid is	*pH = 1*	
Concentrated acids	*have several moles of acid per dm³ of solution*	
The pH of 0.1 M ethanoic acid is	*around pH = 3*	**(6)**

2 pH is a measure of the hydrogen ion concentration. pH = $-\log_{10}[H^+]$. (Note that the higher the concentration of hydrogen ions in a litre of solution, the lower the pH.)   **(2)**

3 a) pH = 13; alkaline; purple.   **(3)**
  b) pH = 4; acid; red.   **(3)**

4 A buffer solution keeps its pH practically constant when small amounts of acid or alkali are added to it.   **(1)**

5 *Type of buffer*	*Use*	*Made from*	
Acidic buffer	Maintains acid pH *i.e.* pH < 7 **(1)**	Weak acid plus the salt of weak acid **(1)**	
Alkaline buffer **(1)**	Maintains pH > 7	Weak alkali plus the salt of the alkali **(2)**	**(5)**

20 Marks   (80% = 16)

5.5.8 Extension – What is the Connection between pH and pK_a?

We have already seen the general equation showing the dissociation of a weak acid:

$$HA(aq) \rightleftharpoons H^+(aq) + A^-(aq) \quad \textit{partially ionised}$$

In a wider interpretation of the nature of acids and bases (Brønsted–Lowry Theory), HA and A^- are said to be conjugate; they form a *conjugate acid–base pair*. Here, HA is the 'conjugate acid' of A^-, and A^- is the 'conjugate base' of HA.

As before: [HA] is the concentration of acid
   $[H^+]$ is the concentration of hydrogen ions
   $[A^-]$ is the concentration of the conjugate base, *i.e.* an ion associated with the acid and capable of recombining with a hydrogen ion.

It is an equilibrium reaction, and the acid dissociation constant can be written as:

$$K_a = \frac{[H^+][A^-]}{[HA]}$$

This can be rearranged to show the concentration of hydrogen ions:

$$[H^+] = K_a \times \frac{[HA]}{[A^-]} = K_a \times \frac{[acid]}{[conjugate\ base]}$$

If we take the logarithm (to base 10) of each side of the expression above, we can rewrite it as:

$$\log[H^+] = \log K_a + \log\left\{\frac{[acid]}{[conjugate\ base]}\right\}$$

Then take negative values of the logarithms:

$$-\log[H^+] = -\log K_a - \log\left\{\frac{[acid]}{[conjugate\ base]}\right\}$$

which is the same as:

$$pH = pK_a - \log\left\{\frac{[acid]}{[conjugate\ base]}\right\}$$

We can use this to calculate the pH of buffer solutions.
   In a buffer solution, comprising a weak acid and its salt:

$$HA(aq) \rightleftharpoons H^+(aq) + A^-(aq) \quad \textit{partially ionised}$$

$$MA(aq) \rightarrow M^+(aq) + A^-(aq) \quad \textit{fully ionised}$$

The majority of A^- ions are supplied by the fully ionised salt, and not by the acid. The concentration of A^- ions will be the same as the concentration of the salt, *i.e.* $[A^-] = [MA]$.

So for a buffer solution, we can rewrite the equation connecting pH and pK_a:

$$pH_{buffer} = pK_a - \log\left\{\frac{[acid]}{[salt]}\right\}$$

Thus, it is possible to estimate the pH of a buffer solution.

Let us look at a special case. *What happens when the buffer contains equal concentrations of acid and salt, i.e.* $[acid] = [salt]$?

• $[acid] = [salt]$, ∴ $[acid]:[salt] = 1$

• $\log\left\{\frac{[acid]}{[salt]}\right\} = \log[1] = 0$

So, for a buffer containing equal concentrations of acid and salt;

• $pH = pK_a$

Likewise, if the concentrations of acid and salt are in the ratio 10:1;

• $[acid]:[salt] = 10:1$

then, $\log\left\{\frac{[acid]}{[salt]}\right\} = \log\left\{\frac{[10]}{[1]}\right\} = 1$

and $pH = pK_a - 1$, the pH has gone down by 1.

• Equally, if $[acid]:[salt] = 1:10$;

• then, $\log\left\{\frac{[acid]}{[salt]}\right\} = \log\left\{\frac{[1]}{[10]}\right\} = -1$

$pH = pK_a - (-1) = pK_a + 1$, the pH has gone up by 1.

Module 6

Compounds
of Carbon (1)

to Sir Harry Kroto, Richard Smalley and Robert Curl . . . The structures are sometimes called Buckyballs! No one yet knows the full potential of these fascinating materials. So carry on with your Chemistry, there is another Nobel prize waiting for someone!!

6.1.5 Compounds of Carbon and Hydrogen – The Hydrocarbons

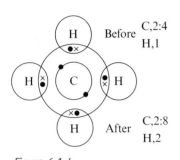
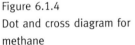

Figure 6.1.4
Dot and cross diagram for methane

Consider methane, CH_4, which is the simplest of the hydrocarbon compounds. It has the four electrons of the outer shell of carbon shared with one electron each from four hydrogen atoms. The electron shell of hydrogen is now full, and so stable, with two electrons – it has the 'helium' outer electron structure. The carbon shell now has eight electrons in it, and so it is also full and stable.

A pair of electrons can be represented by the straight lines shown between the carbon and the hydrogen.

The molecule is not flat but is a three-dimensional structure in space and the C–H bonds are all pointing away from the carbon towards the corners of a tetrahedron (a triangular pyramid). The angle between each C–H and the next one is 109° 28' (or approximately 109.5°). Because it is sometimes difficult to show these angles on paper the structures are often written as though they were flat. Use a molecular model kit to show the real three-dimensional structure. Some of the ways of displaying these structures are included in Figure 6.1.5, showing the various ways of trying to represent three-dimensional structures on two-dimensional paper.

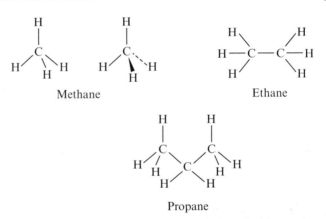

Figure 6.1.5
Representations of hydrocarbon structures

All of the straight lines between the atoms represent pairs of electrons forming covalent 'overlap' bonds which are called sigma (σ) bonds. All of the hydrocarbons mentioned so far are 'saturated' molecules, meaning that there are no double bonds present in the molecule (see also Section 2.2.2).

Single bonds (sigma bonds) will also be seen in the carbon–oxygen bonds discussed later in the alcohol series, where there is a C–O–H bond. Double bonds (sigma and pi bonds) also occur in aldehydes and ketones where there is a C=O bond. These homologous series will be looked at in Module 7.

Carbon can also form 'unsaturated' carbon–carbon double bonds in molecules, *e.g.* as in ethene. The second bond is called a pi (π) bond and forms above and below the sigma bond. Molecules containing double or even triple bonds are called 'unsaturated' (Figure 6.1.6) (see also Section 2.2.4).

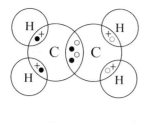

Every ring has either the helium structure of two electrons or the stable octet of eight electrons

Ethene

The H—C—H bonds here are all 120 degrees to each other, and the molecule is flat. Each pair of electrons is shown by a straight line

Figure 6.1.6

Bonding in ethene

Other homologous series which we shall meet, together with their general names and formulae, are outlined in Table 6.1.1.

Table 6.1.1 *Some common homologous series*

Class	Functional Gp	General Formulae	Prefix	Suffix	Example
Alcohols	–OH, hydroxy	$C_nH_{2n+1}OH$	hydroxy-	-ol	C_2H_5OH, ethanol
Amines	–NH$_2$, amino	$C_nH_{2n+1}NH_2$	amino-	-amine	$C_2H_5NH_2$, ethylamine
Amides	–CONH$_2$	$C_nH_{2n+1}CONH_2$		-amide	$C_2H_5CONH_2$, propanamide
Aldehydes	–CHO	$C_nH_{2n+1}CHO$		-al	C_2H_5CHO, propanal
Halogenoalkanes or alkyl halides	–Cl, –Br, –I	$C_nH_{2n+1}X$	chloro-, bromo-, etc.		C_2H_5Cl, chloroethane
Carboxylic acids	–COOH carboxyl	$C_nH_{2n+1}COOH$		-oic acid	CH_3COOH, ethanoic acid
Esters	–COOR	$C_nH_{2n+1}COOR$		alkyloate	$C_2H_5COOC_2H_5$, ethyl propanoate
Ethers	–O–	C_nH_{2n+1}–O–C_nH_{2n+1}	alkoxy-	ether	$C_2H_5O\,C_2H_5$, ethoxyethane or diethyl ether
Ketones	–C=O	$C_nH_{2n+1}CO\,C_nH_{2n+1}$	oxo-	-one	CH_3COCH_3, propanone

6.1.6 Summary

- Carbon compounds are very abundant and often come from natural materials.
- All carbon compounds are covalent.
- Carbon can form single σ bonds or double σ and π bonds.
- Carbon can form covalent bonds with other elements including hydrogen, oxygen, nitrogen and the halogens.
- Series of organic compounds are classified by their functional groups.

ALIPHATIC AND AROMATIC

When discussing organic compounds, the term 'aliphatic' is used to mean open chains of carbons in a molecule (Figure 6.1.7), and 'aromatic' means that the compound contains alternate single and double bonds in an unsaturated ring of carbon atoms, resulting in delocalisation of electrons as in benzene, C_6H_6 (Figure 6.1.8). (A more detailed study of aromatic compounds occurs in Unit 6.3.)

Figure 6.1.7

'Aliphatic' or open chain structure

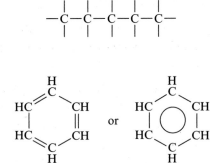

Figure 6.1.8

'Aromatic' structure

Answers to diagnostic test

If you score less than 80%, then work through the text and re-test yourself at the end using this same test. If you still get a low score then re-work the topic at a later date.

1 C–H bonds are <u>covalent and stable</u>, each atom having a <u>complete shell</u>. Carbon can form <u>strong C–C bonds</u> easily, and so there are many opportunities for these to form in <u>chains of various lengths (or in rings)</u>.   (4)

2 <u>Living things, plants and animals</u> when they die are the source of <u>organic compounds</u>. Ultimately, all the carbon in plants comes from the carbon dioxide in the air.   (2)

3 A series of compounds all containing the <u>same kinds of atoms</u> and all the compounds having the same <u>overall general formula</u>, *e.g.* alkanes C_nH_{2n+2}.   (2)

4 Because C is in <u>Group 4 of the periodic table</u> and gains its stable octet only by <u>sharing electrons</u>, *i.e.* covalent bonding.   (2)

5 In alkanes all the bonds are <u>saturated sigma bonds</u>, whereas in alkenes the carbon to carbon bond is <u>unsaturated</u>. The carbon-to-carbon double bond in an alkene consists of a sigma bond together with a pi bond, as seen in Figure 6.1.6.   (2)

6 Because organic molecules are covalent then there is very <u>little attraction between one molecule and its neighbour</u>, and so it only takes a <u>low temperature</u>, say body temperature, to separate the molecules. So the liquid boils at a low temperature, it <u>evaporates easily</u> and you smell it.   (3)

15 Marks   (80% = 12)

FURTHER QUESTIONS (NO ANSWERS GIVEN)

1 In about 500 words give an account of the impact that organic chemistry has upon our lives.

2 Oils and coal are great sources of organic compounds and it is a waste merely to burn them as a fuel. Discuss this with specific examples in about 200 words.

3 Organic compounds often have characteristic odours and are volatile. This is due to the physical properties of these compounds. Discuss this in about 200 words.

Unit 6.2
Compounds of Carbon and Hydrogen – Alkanes and Alkenes

Starter check

What you need to know before you start this unit.

You should already know about the characteristics and general properties of covalent bonds, something about bond angles, sigma and pi bonds (Unit 2.2), and what is meant by the term 'homologous series' (Unit 6.1).

Aims **By the end of this unit you should understand:**

- The chemistry of some carbon and hydrogen compounds including the system of naming; physical properties of hydrocarbons.
- The structures, typical reactions, properties and uses of the alkane and alkene series, including substitution and addition reactions.

Diagnostic test

Try this test at the start of the unit. If you score more than 80%, then use this unit as a revision for yourself and scan through the text. If you score less than 80%, then work through the text and re-test yourself at the end by using this same test.

The answers are at the end of the unit.

1 What are the main differences between the structures of ethene and ethane? (2)
2 Draw the structures of 2,3-dimethylbutane and 2,3-dimethylpent-2-ene. (2)
3 Draw the structures of the first (C2) and third (C4) members of the monoalkene series. (4)
4 a) In what way do alkanes and alkenes react with chlorine gas? Give equations using the first member of each series as the example. (6)
   b) Which hydrocarbon series would you expect to polymerise? Give an example. (2)
   c) What is meant by cracking of a saturated hydrocarbon? (2)

5 Write the balanced equation for the burning of butane in
oxygen.   (2)

20 Marks   (80% = 16)

6.2.1 Alkanes

The homologous series of alkanes has the general formula C_nH_{2n+2}.

The importance of small and medium-sized alkane molecules to society is that they are excellent fuels for cars, planes, boats and homes. The smell of petrol at the petrol station is due to the mixture of covalent hydrocarbon compounds.

The alkanes are very abundant in the natural gas and crude oils found under the North Sea (Figure 6.2.1), in the Middle East, USA and many other areas throughout the world.

Figure 6.2.1
Oil rigs are used in the extraction of gas and oils from beneath the North Sea

The thick, black crude oil is a mixture of thousands of hydrocarbons. The mixture can be separated into smaller 'fractions', each containing molecules of similar boiling points, by distillation. The fractions with higher boiling points obviously contain larger and heavier hydrocarbon molecules. These can be broken up into smaller molecules in the oil refineries and processing plants. The breaking up of big oily molecules into the smaller more useful ones is done by 'cracking' (Figure 6.2.2). The smaller molecules are more useful for making easily combustible fuels and for making plastics and other useful chemicals. This is why its such a waste to lose them by burning them as cheap fuels.

$$CH_3-CH_2-CH_2-CH_2-CH_2-CH_2-CH_2-CH_2-CH_2-CH_2-CH_2 \ldots etc.$$

$$\downarrow \text{Heat + Catalyst}$$

$$CH_2=CH_2 \; + \; CH_3-CH_2-CH_3 \; + \; CH_4 \; + \; CH_3-CH_3 \; etc.$$
alkenes and alkanes

Figure 6.2.2
'Cracking' of long chain
hydrocarbon molecules

Catalytic cracking, using heat and a catalyst, breaks up the long molecules – which are sometimes over a hundred carbon atoms in length – into smaller and more useful molecules, and these can be separated by fractional distillation because each material will have its own characteristic boiling point (Figure 6.2.3). Notice in Figure 6.2.2 that cracking an alkane will always produce some alkenes.

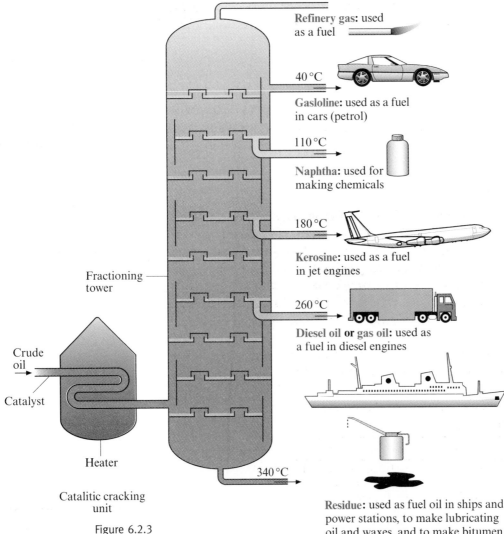

Fractioning tower

Crude oil

Catalyst

Heater

Catalitic cracking unit

Figure 6.2.3
Fractional distillation of crude oil

Refinery gas: used as a fuel

40 °C

Gasloline: used as a fuel in cars (petrol)

110 °C

Naphtha: used for making chemicals

180 °C

Kerosine: used as a fuel in jet engines

260 °C

Diesel oil or gas oil: used as a fuel in diesel engines

340 °C

Residue: used as fuel oil in ships and power stations, to make lubricating oil and waxes, and to make bitumen for surfacing roads

There is sometimes a conflict between what society demands and what chemists are able to deliver. For example, perhaps the bonanza of North Sea oil would have gone further if the organic chemists had had their way, rather than the economists who encouraged the government to make a 'quick buck' selling off the oil as fuels. A lot of valuable and irreplaceable chemicals were lost by burning up our coal, oil and natural gas rather than using more of it to synthesise polymers and fibres. The hydrocarbons have been burned into carbon dioxide gas and water and all the special properties of the useful materials lost for ever. Natural gas is largely methane, and most of it is burnt to release heat for central heating, cooking and in power stations:

$$CH_4(g) + 2O_2(g) \rightarrow CO_2(g) + 2H_2O(l) + \text{heat given out}$$
(exothermic reaction)

Excess carbon dioxide is thought to be one of the causes of the 'greenhouse effect' and 'global warming' (see Unit 7.4). Burning up too much fossil fuel can have an adverse affect on our atmosphere and environment.

Society demands so much energy, and fuel for transport, that chemists have been called in to try to solve the problems. Some of the solutions will not be popular! In the past, thousands of tons of coal were burned daily in Britain, but how many coal mines are open now? We use millions of litres of oil and millions of cubic metres of gas daily, but for how long? The processes by which crude oil and natural gas are formed are far too slow for replacements to be made for millions of years. What alternative fuels are there?

Let's get back to the organic chemistry.

The simplest organic compounds are those which contain only carbon and hydrogen atoms – these are called hydrocarbons. All the bonds are covalent.

NAMING SYSTEM

Alkanes are covalent compounds in an homologous series which has the general molecular formula: C_nH_{2n+2}. Thus for methane, which is the simplest of the series, $n = 1$ (there is only one carbon atom), so there are $(2 \times 1) + 2 = 4$ hydrogen atoms; hence methane's formula is CH_4.

Any compound that has only one carbon atom in it has the prefix **meth** in its name. Any saturated hydrocarbon with a general formula C_nH_{2n+2} will have a name ending in **ane**. Hence CH_3H would be **meth . . . ane**, or **methane**.

The first six members of the alkane series are shown in Table 6.2.1. You can fill in the missing formulae and names in the table.

Table 6.2.1 *The first six members of the alkane homologous series* (C_nH_{2n+2})

n	Formula	Name	Name of radical
1	CH_4	Methane	CH_3-, methyl or meth-
2	C_2H_6	Ethane	C_2H_5-, ethyl or eth-
3	C_3H_8	Propane	C_3H_7-, propyl or prop-
	You write their formulae for the rest		
4		Butane	
5		Pentane	
6		Hexane	

The first four members of many homologous series often have 'trival' or non-systematic names, many dating back to the early history of chemistry, but from then on they have more systematic names which show how many carbon atoms there are present; *e.g.* $n = 5$ is **pentane**, ('pent' meaning five in Greek). 'Hex' means six in Greek, so $n = 6$ = **hexane**.

The **empirical formula** of a compound shows the simplest ratio of each kind of atom in a molecule of the compound, *e.g.* CH_3 for ethane.

Table 6.2.4. These values have been plotted on a suitable scale and Figure 6.2.10 shows the general trends.

Add the names and structures of the alkanes to the table.

Table 6.2.3 *Melting and boiling points for alkanes*

For straight chain alkanes only		No. of C atoms (n)	Melting and boiling points/K	
name	structure		mp	bp
		1	91.1	109
		2	89.9	185
		3	83.4	231
		4	135	273
		5	143	309
		6	178	342
		7	183	372
		8	216	399
		9	222	424
		10	243	447

Now fill in the structures of the named alkenes and interpret the meaning of the graphs.

Table 6.2.4 *Melting and boiling points for alkenes*

For straight chain alkenes only		No. of C atoms (n)	Melting and boiling points/K	
name	structure		mp	bp
Ethene		2	104	169
Propene		3	88	225
But-1-ene		4	88	276
Pent-1-ene		5	108	302
Hex-1-ene		6	133	336

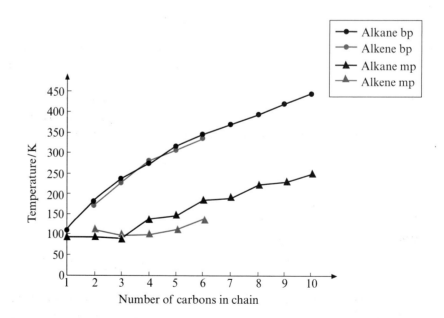

Figure 6.2.10
Melting and boiling points of
some simple hydrocarbons

The boiling points of the alkanes and alkenes follow a similar pattern, but the melting points are slightly different. Can you offer any possible explanation? You might like to compare the melting and boiling points of the branched chain compounds of the alkanes and offer an explanation. Consider the 2-methyl alkane derivatives in Table 6.2.5. The values for the straight chain compounds containing the same number of carbons were listed in Table 6.2.3. Do you think the branched molecules fit more easily into the solid crystal form and that these molecules are held more strongly to each other than the straight chain compounds? Or not? Plot the values from Table 6.2.5 on to the same graph as Figure 6.2.10 to see whether there are any general patterns to be drawn out. Interpret the meaning of the shape of the graph.

Table 6.2.5 *Melting and boiling points for simple branched chain alkanes*

No. of C atoms	Name	mp/K	bp/K
4	2-methylpropane	114	261
5	2-methylbutane	113	301
6	2-methylpentane	119	333
7	2-methylhexane	155	363
8	2-methylheptane	164	390

WHAT IS MEANT BY SATURATED AND UNSATURATED CARBON COMPOUNDS?

Alkenes are called 'unsaturated' because they contain a C=C double bond and therefore less hydrogen than the corresponding alkanes. They can undergo addition reactions in order to become 'saturated'. The alkanes have no double bonds and all their bonds are sigma bonds between C–C and C–H. They cannot undergo addition reactions because they are saturated already. An even more unsaturated series of hydrocarbons is the alkynes, C_nH_{2n-2}, which contain a carbon–carbon triple bond, (C≡C), made up of a sigma and two pi bonds. The best known member of this series is C_2H_2, ethyne (also known as acetylene).

HYDROCARBONS AS FUELS . . . COMBUSTION

All hydrocarbons will burn in air or oxygen to produce carbon dioxide and water. All give out heat when they burn, *i.e.* the reaction is exothermic.

The reaction requires a spark or a flame to start the burning, but once started it continues on its own and gives out heat. So hydrocarbons (particularly the lighter members of the series) are all potentially good fuels.

Methane: $CH_4(g) + 2O_2(g) \rightarrow CO_2(g) + 2H_2O(l)$

Ethane: $2C_2H_6(g) + 7O_2(g) \rightarrow 4CO_2(g) + 6H_2O(l)$

Ethene: $C_2H_4(g) + 3O_2(g) \rightarrow 2CO_2(g) + 2H_2O(l)$

General equation: $C_xH_y(g) + (x + y/4)O_2(g) \rightarrow xCO_2(g) + (y/2)H_2O(l)$

You will be aware that North Sea gas consists mainly of methane. Propane and butane are used as bottled gas, camping gas or in cigarette lighters (when under pressure both propane and butane are liquids). Because the burning process gives out heat, the 'exothermic' reaction is put to commercial use:

6.2.4 Extension – Some Explanations and Mechanisms

'Mechanism' means an explanation of the individual detailed steps occurring when a reaction takes place.

THE DETAILED MECHANISM FOR THE REACTION OF CHLORINE WITH METHANE IN BRIGHT SUNLIGHT

This reaction was one of the first very fast or explosive reactions to be studied in detail by research chemists. The reaction makes use of a very short-lived species called a **free radical**. These are particles with an odd number of electrons associated with them, made by the symmetrical splitting of a covalent bond ($Cl_2 \rightarrow Cl* + Cl*$); hence they are very reactive as they have a great tendency to link up with another electron from other molecules. The mechanism goes in three very fast steps: **initiation**, when the free radicals are formed, **propagation** of the reaction by producing the various products and more of the reactive radical species, and then **termination** when one of the reactants runs out or if the free radicals join together.

It is worth noting that the energy given out when the two free radicals join together is exactly equal to the energy needed to break the covalent bond apart, so the bond cannot be formed unless something is present to take away the excess energy. This is why the free radical chain reactions can sometimes go on a long time before they stop or if something is added to stop them, as seen in Figure 6.2.23.

ALKENE + HBr MECHANISM

This is the more detailed mechanism of what happens when an addition reaction takes place between an alkene and HBr.

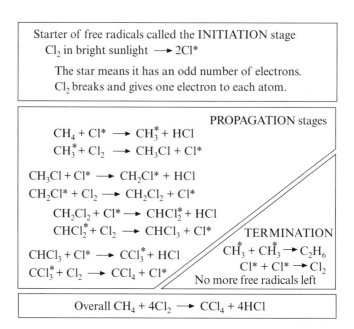

Figure 6.2.23
A free radical chain mechanism

Ethene gas when mixed with HBr gas produces droplets of a colourless liquid, bromoethane (Figure 6.2.24).

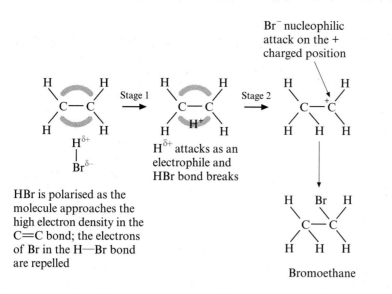

Figure 6.2.24

Mechanism for the formation of bromethane

As discussed earlier, if propene gas and HBr gas are mixed then the H of the H–Br attacks the carbon atom with the most number of hydrogens on it and the Br goes to the other carbon atom.

These two alternative reactions (Figure 6.2.25) are only possible in unsymmetrical alkenes.

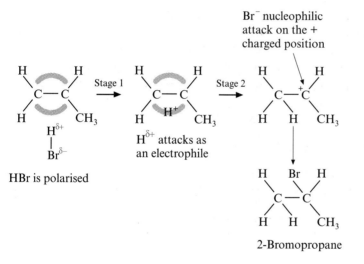

The Kharasch mechanism would produce 1-bromopropane by a slightly different mechanism involving free radicals and is not discussed here

Figure 6.2.25

Formation of bromopropane

Unit 6.3
Aromatic Organic Chemistry

Starter check

What you need to know before you start this unit.

You should already know about open chain homologous series, covalent bonding including sigma and pi bonds, and also know what is meant by the terms 'aromatic' and 'substitution'.

Aims **By the end of this unit you should understand:**

- The chemistry of some aromatic compounds.
- The structures, typical reactions, properties and uses of the arenes, including substitution reactions.

Diagnostic test

Try this test at the start of the unit. If you score more than 80%, then use this unit as a revision for yourself and scan through the text. If you score less than 80%, then work through the text and re-test yourself at the end by using this same test.

The answers are at the end of the unit.

1 How is the structure of benzene, C_6H_6, different from a six-carbon alkane or alkene? **(3)**

2 Draw the structure of benzene, and also the three different structures for dimethylbenzene. **(4)**

3 What would be the systematic names of the compounds of dimethylbenzene drawn in Question 2? **(3)**

4 What is the usual type of reaction undergone by the aromatic benzene ring, substitution or addition? Show how benzene reacts with a mixture of concentrated nitric and sulfuric acids. **(2)**

5 Show the structure of benzene, with the pi-cloud of electrons. **(2)**

6 Choose two examples to show how different groups on the benzene ring influence the position of a second group entering the ring. **(4)**

18 Marks   (80% = 14)

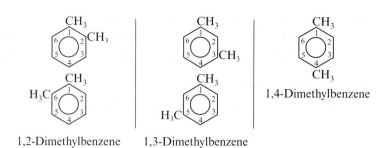

Figure 6.3.4
Possible isomers of
dimethylbenzene

1,2-Dimethylbenzene   1,3-Dimethylbenzene

1,4-Dimethylbenzene

6.3.3 Other Compounds of Benzene

Other typical compounds may be formed by benzene in **substitution** reactions where one or more hydrogen atoms are replaced (or substituted) by other atoms or groups. Three examples are given in Figure 6.3.5.

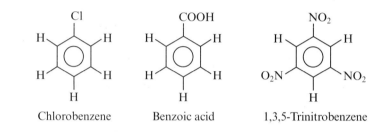

Figure 6.3.5
Common substitution
products of benzene

Chlorobenzene   Benzoic acid   1,3,5-Trinitrobenzene

We have already explained that where **two** hydrogen atoms are substituted, there are, therefore, three different possible arrangements of substituents on the ring. If the arrangements are different, then the compounds are different, with slightly different physical properties, *etc*. They are named according to their positions relative to each other, as in Figure 6.3.6.

The first group X is on position 1

Figure 6.3.6
Naming compounds according to
the position of the substituent

The second group can enter at position 2 or 3 or 4. Positions 5 and 6 are the same as the 3 and 2 positions, respectively. It is always important to give numbers to the positions.

Look at the possible arrangements of the di-substituted chloro-compounds of benzene (Figure 6.3.7). What would the compounds in the bottom half of the figure be called? . . .Think first!

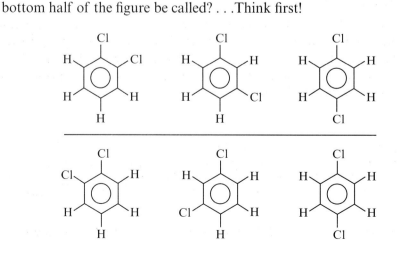

Figure 6.3.7
Di-substituted benzenes

Can you see that these will have the same names as the previous ones? This is because although the structures are viewed from the opposite side, they are the same compounds. Models of these compounds will show this.

The chemistry of aromatic compounds will not be pursued further in this text except for a few selected reactions of benzene.

6.3.4 Substitution Reactions

As mentioned previously, although the benzene ring is 'unsaturated' it is generally very stable and does not give the typical addition reactions of 'unsaturated' compounds. Its reactions are generally substitution reactions.

Consider the reaction between benzene and nitric acid (in the presence of concentrated sulfuric acid) with the mixture kept below 50 °C (Figure 6.3.8).

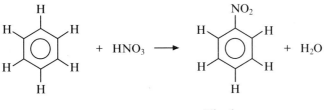

Nitrobenzene

Figure 6.3.8
Nitration of benzene

Another way of showing the same thing is given in Figure 6.3.9.

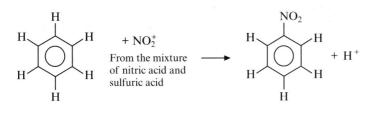

Figure 6.3.9
Nitration of benzene

$$H_2SO_4 + HNO_3 \rightarrow H_2NO_3^+HSO_4^-$$
$$H_2NO_3^+ \rightarrow NO_2^+ + H_2O$$

The concentrated sulfuric acid helps to make the ions, NO_2^+, needed for the reaction. This positive ion is attracted to the high electron density in the benzene ring. Notice how one hydrogen was replaced by a nitro group.

If further nitric acid is added and the reaction warmed above 50 °C a further nitro group is added to the ring. It enters at the 3-position forming 1,3-dinitrobenzene. Further substitution or nitration is possible, but more heat is needed to form 1,3,5-trinitrobenzene (Figure 6.3.10). Notice that the NO_2 makes the ring less active and to add a further NO_2 requires stronger conditions.

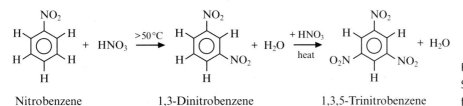

Nitrobenzene   1,3-Dinitrobenzene   1,3,5-Trinitrobenzene

Figure 6.3.10
Stronger conditions bring about further nitration

6 Draw the structures and explain the isomerism, if any, occurring in the following:

   a) but-2-ene
   b) but-1-ene
   c) propane
   d) propan-2-ol
   e) propan-1-ol
   f) dimethylbenzene   **(9)**

7 Complete the sentences:

   a) Two geometric isomers differ in their _____ properties but have similar _____ properties.
   b) The two isomers of C_2H_6O have different _____ properties and different _____ properties.
   c) The three ring isomers of dimethylbenzene have different _____ properties and give _____ products on nitration with nitric acid.
   d) The isomers of butane differ in their _____ properties.   **(7)**

32 Marks   (80% = 26)

6.4.1 Isomerism

When studying organic chemistry it becomes apparent that some molecular formulae, often with more than three carbon atoms, have alternative structures. This phenomenon is called **isomerism**.

Isomers are structures that have the following characteristics:

a) the same molecular formula;
b) the same molecular mass;
c) different structures or sequences of atoms linked together;
d) different physical properties and often separate distinct chemical properties;
e) different systematic names.

As the number of atoms in the compound increases within a homologous series, so the number of isomers increase rapidly. So butane has 2 isomers, pentane has 3, hexane has 5, heptane has 9 and decane has 75.

It is necessary to use a molecular model kit to give a three-dimensional and realistic representation of the actual appearance of the shapes of molecules when considering isomerism.

For convenience, this large topic will be studied under the following headings:

Set 1 Structural Isomerism
• Structural Isomerism in Aliphatic Hydrocarbons:
  a) alkanes;
  b) substituted alkanes;
  c) alkenes – the position of the double bond.
• Functional Group Isomerism:
  a) alcohols and ethers;
  b) aldehydes and ketones;
  c) carboxylic acids and esters.
  d) isomerism in aromatic compounds.
Set 2 Stereo-isomerism
• Optical Isomerism
  a) alkanes;
  b) substituted alkanes.
• Geometric or *cis–trans* isomerism.

6.4.2 Set 1 – Structural Isomerism

STRUCTURAL ISOMERISM IN HYDROCARBONS

Alkanes

Because methane has a tetrahedral shape, with the H–C–H bond angles 109° 28', the more accurate method of showing the shapes and bond angles of the alkanes is shown in Figure 6.4.1.

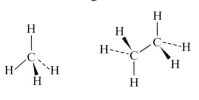

Methane   Ethane

Figure 6.4.1

3-D representations

Alcohols are isomeric with ethers ($C_nH_{2n+2}O$), but they have very different chemical reactions.

Aldehydes and ketones

Similarly, aldehydes and ketones can have the same overall molecular formula but different structures. Each series contains a carbonyl group (C=O) but in each compound they are attached to different groups. The ketone has always got a C–CO–C chain, whereas the aldehyde has a CHO group which is always at the end of the molecule. The first member of the ketone series, propanone, is isomeric with the aldehyde propanal (Figure 6.4.11).

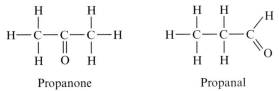

Figure 6.4.11
Aldehyde and ketone

Propanone   Propanal

Carboxylic acids and esters

The two functional groups here are the carboxylic acid and ester groups. Any carboxylic acid will have a corresponding ester isomer. A carboxylic acid contains the COOH group, whereas the ester always has a COOR group, where R can be any aliphatic or aromatic group. So ethanoic acid (as in vinegar) is isomeric with methyl methanoate (Figure 6.4.12).

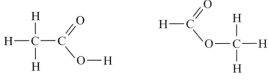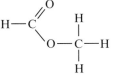

Figure 6.4.12
Carboxylic acid and ester

Ethanoic acid
(or 'acetic acid')

Methyl methanoate

Isomerism in aromatic compounds (a sub-group of structural isomerism)

The aromatic system to be considered will be the six-membered rings of benzene as seen in the lower part of Figure 6.3.3 on page 311.

Isomerism in substituted benzene rings needs two or more groups attached to the ring, and depends on the relative positions of the groups on the ring. Isomers of this type have already been met in Unit 6.3.

All of the positions on the ring are identical, so if there is no group or only one group joined to the ring there is only one possible structure.

We know that the typical compounds formed by benzene are **substitution** products in which one or more hydrogen atoms are replaced (or substituted) by other atoms or groups (three examples are given in Figure 6.4.13).

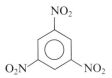

Figure 6.4.13
Aromatic compounds

Chlorobenzene

2-Hydroxybenzoic acid (or 'salicylic acid') used to make aspirin

1,3,5-Trinitrobenzene

There are a three different arrangements on the ring that **two** substituents can form. They are named according to their positions relative to each other (look back at Figure 6.3.6 on page 312).

Look at the possible isomers of the di-substituted chloro compounds of benzene in Figure 6.1.14. What would the following compounds be called? Think first! The answers are given in Figure 6.1.15.

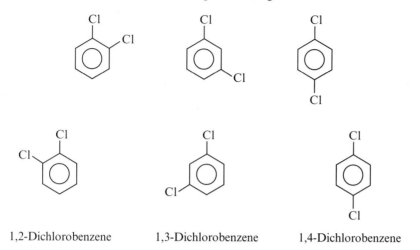

Figure 6.4.14
Di-substituted chlorobenzenes

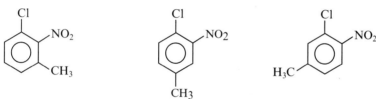

1,2-Dichlorobenzene   1,3-Dichlorobenzene   1,4-Dichlorobenzene

Figure 6.4.15
Do they have different names
from the compounds above?

Can you see that these will have the same names as the previous ones? The only thing that matters is the relative position of the two groups.

Figure 6.4.16 shows isomers of chloronitromethylbenzene. If there are three or more substituents on the ring, the principle of naming is still the same.

Figure 6.4.16
Chloronitromethylbenzenes

3-Chloro-2-nitromethylbenzene   4-Chloro-3-nitromethylbenzene   3-Chloro-4-nitromethylbenzene

6.4.3 Set 2 – Stereo-isomerism

OPTICAL ISOMERISM

Optical isomerism occurs where there are four different atoms or groups of atoms joined to a central carbon atom. There are two possible arrangements, and each will be a mirror image of the other. The isomer structures will not be superimposable one on the other, and each will interact with the plane of polarised light, twisting the light in either a clockwise or anticlockwise direction.

We say that such molecular structures are **asymmetric**, and it is the asymmetry around the carbon atoms which enables the formation of a pair of **optically active** isomers. The isomers are mirror images of each other (Figure 6.4.17).

The two isomers are different, and a bit like your hands; if you put your hands flat together they are mirror images of each other (Figure 6.4.18). You cannot put your hands one on top of the other so that they match thumb on top of thumb. Neither can you place optical isomers one on top of the other. Make the models and try it. They are non-superimposible.

In butene it is possible to form **isomers** according to both the location of the double bond (structural isomerism) and also the geometric arrangement around a double bond. These are shown in Figure 6.4.24.

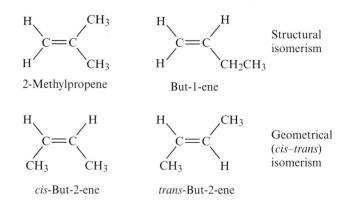

2-Methylpropene   But-1-ene

Structural isomerism

cis-But-2-ene   trans-But-2-ene

Geometrical (*cis–trans*) isomerism

Figure 6.4.24
Butenes exhibit structural and geometrical isomerism

But-2-ene exhibits **geometrical isomerism** or ***cis–trans* isomerism**.

cis–trans Isomerism is another form of stereo-isomerism. The sequence of atoms joined together in the two isomers is exactly the same; it is their arrangements in space which are different.

The characteristics of this type of isomerism are that:

• There must be restricted rotation around a double bond.
• The two isomers contain the same atoms in the same sequence but with different arrangements in space.
• Neither C atom can have identical atoms or groups attached to it. If one does, *cis–trans* isomerism is not possible.
• The isomers have different melting points and other physical properties.

Another example is 1,2-dibromoethene, as in Figure 6.4.25.

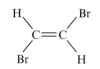

Figure 6.4.25   cis-1,2-Dibromoethene   trans-1,2-Dibromoethene

Answers to diagnostic test

If you score less than 80%, then work through the text and re-test yourself at the end by using this same test. If you still get a low score then re-work the topic at a later date.

1 In each part give one mark for a satisfactory definition and one for a diagram.

a) 'Structural' isomerism. This occurs when for the same molecular formula there are at least two different structures. (The same atoms are connected together in a different sequence.)
The structural isomers of butane are as follows:

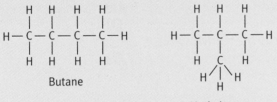

Butane   2-Methylpropane

b) Geometrical isomerism. This is a form of stereo-isomerism, which usually involves compounds containing C=C bonds. The isomers differ only in the way the groups are distributed on either side of the double bond. The isomers of but-2-ene are:

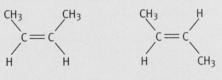

cis-But-2-ene   *trans*-But-2-ene

c) Optical isomerism is another form of stereo-isomerism in which the two isomers have the same structural formula but differ only in their orientation in space. They are related to each other as non-superimposable mirror images. (Four different atoms or groups attached to a central carbon atom.) **(6)**

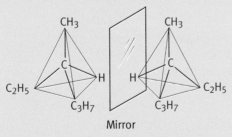

Mirror

2 Structural isomerism occurs in butane. Its structures are as above, in Question 1(a). **(3)**

3 Geometrical isomerism occurs in but-2-ene. Its structures are as in Question 1(b). **(3)**

4 Structural isomerism exists in dichlorobenzene. **(1)**

7.1.1 Introduction

Alcohols are best known for the single member of the series which is known as 'alcohol', although it should be known as 'ethanol'.

The physiological effects of the alcohols are not all the same. The first member of the series, methanol (or methyl alcohol), gives some initial stimulation, but excess can cause 'drunkenness' and a loss of co-ordination. More dramatically, it has the side-effect of damaging the optic nerve, causing the drinker of this 'alcohol' to go blind. It eventually damages other nerves and causes insanity and death. Methylated spirits ('meths'), is ethanol deliberately 'denatured' by mixing with methanol. Drinking this is definitely not recommended!

Ethanol, the second member of the series (and the one found in alcoholic drinks), causes, in the diluted state in which it is found in drinks, the alcoholic effects of initial stimulation which are often followed by depression. As the quantity taken is increased, it causes loss of control of speech and temporary loss of muscle control, the nervous system is affected and control of mind and limbs is lost. The rich sweet smell of ethanol is also well known and characteristic.

The higher alcohols are much more toxic and distasteful. They are very useful as solvents for organic compounds, and as starting points for making compounds in other homologous series. They are certainly not drinkable and many are poisonous.

The general formula of the alcohols series is $C_nH_{2n+1}OH$. For ethanol, $n = 2$, and the structure is shown in Figure 7.1.1.

Their names are obtained by simply changing the 'e' of the 'ane' of alkane with the same number of carbon atoms to 'ol'. Thus, the early members of this homologous series are given in Table 7.1.1

It has been shown that, for many people, alcohol is an addictive drug, with psychiatric disturbance associated with physiological changes in the brain. Serious alcoholism causes physical shrinkage of the brain within the skull.

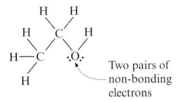

Two pairs of non-bonding electrons

Figure 7.1.1
Ethanol

Table 7.1.1 *Homologous series of simple alcohols*

n	Name	Molecular formula	Relative Molecular mass	Boiling point/K	Structures
1	Methanol	CH_3OH	32	338	Draw these
2	Ethanol	C_2H_5OH	46	352	structures
3	Propan-1-ol	C_3H_7OH	60	370	yourself
3	Propan-2-ol	C_3H_7OH	60	356	
4	Butan-1-ol	C_4H_9OH	74	390	
4	Butan-2-ol	C_4H_9OH	74	373	
4	2-Methylpropan-2-ol	C_4H_9OH	74	355	
5	Pentan-1-ol	$C_5H_{11}OH$	88	411	
6	Hexan-1-ol	$C_6H_{13}OH$	102	431	

TO DO . . .

Draw a graph of boiling points, along the vertical axis, against the value of 'n' on the horizontal axis. Note the shape. Draw a line through the values for the following, methanol, ethanol, propan-1-ol, butan-1-ol, pentan-1-ol, hexan-1-ol. Can you explain the shape of the graph in terms of the structures of these alcohols?

What can be said about the reasons why the boiling points of the three C_4H_9OH structures are different? Look at the different shapes of the molecules.

The boiling points, and hence the physical state, of the lower members at room temperatures could not be predicted by looking at just the relative molecular masses. Usually, the higher the relative molecular mass (RMM) of a compound the higher the boiling point. For the alkanes, the lower members (the first four) are gases, the next few are low boiling liquids, and the higher ones are waxy solids. Methanol and ethanol, the first and second members of the alcohol series (RMMs 32, 46) are liquids, whereas propane, the third member of the alkane series with RMM of 44 is still a gas. The extra energy needed to separate one molecule of ethanol from another to vapourise it caused by the same phenomenon that keeps water molecules held together in groups of threes or fours – hydrogen bonding (see Figure 7.1.2). (An explanation of why this type of bonding is possible is included in Section 2.3.4.)

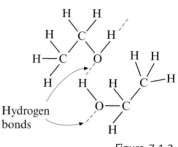

Hydrogen bonds

Figure 7.1.2

You might also want to refer to Section 6.2.3 and compare the boiling points of the alkanes with alcohols containing the same number of carbon atoms. This will help show more clearly the effect of hydrogen bonding on the boiling points of the alcohols as compared with alkanes, which have no hydrogen bonding. If there was no hydrogen bonding present, then the boiling points of the alcohols would steadily increase due to their higher RMMs and the weak van der Waals' forces holding the molecules together. These forces act between **all** particles, and are caused by small oscillations in electron density causing momentary polarity, which induces opposite polarity in any adjacent particles, so leading to a very weak electrostatic attraction. The approximate relative strengths are: van der Waals' attraction, 1; hydrogen bond, 10; covalent bond, 100.

7.1.2 Types of Alcohols

The alcohols in Table 7.1.1 are all monohydric, *i.e.* they contain only one hydroxy (OH) group per molecule. There are homologous series that contain two (dihydric) or three (trihydric) OH groups but these are beyond the scope of this section. The properties of the OH groups in these compounds are similar to the OH groups in monohydric alcohols. Just in passing, the dihydric alcohol, ethane-1,2-diol, is used as an antifreeze in cars, and propane-1,2,3-triol is known as glycerol or glycerine, which is sometimes used in cosmetics and as a sweetener in some food products (Figure 7.1.3). It can also be 'nitrated' to form the high explosive, nitroglycerine!

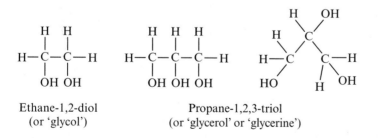

Ethane-1,2-diol
(or 'glycol')

Propane-1,2,3-triol
(or 'glycerol' or 'glycerine')

Figure 7.1.3

Isomerism occurs in alcohols as it does in other organic compounds. A brief account of isomerism is given here but a fuller account has already been given in Unit 6.4. There are two possible structural formulae for propanol, shown in Figure 7.1.4 (use a model kit to make these).

In propan-1-ol the 'OH' group is attached to a C atom which has two H atoms on it, *i.e.* it contains a CH$_2$OH group. This is known as a **primary alcohol**.

In propan-2-ol the 'OH' group is attached to a C atom which itself is

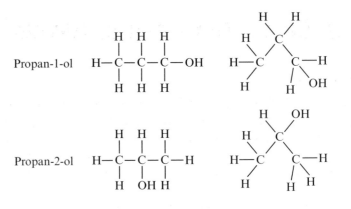

Figure 7.1.4

Structural formulae for propanol

Propan-1-ol

Propan-2-ol

attached to two other C atoms and so contains the $\diagdown$CHOH group. This is known as a **secondary alcohol**.

These have the same molecular formula but different structural formulae – so they are isomers (see Section 6.4.2). The only difference between them is that the 'OH' group is in a different position in each molecule. Both structures have the same molecular formula C_3H_7OH.

Propanol does not have a tertiary alcohol isomer. The first of the alcohol series to have these isomers is when $n = 4$, C_4H_9OH.

Tertiary alcohols have an OH joined to a carbon atom which is attached to three other carbon atoms; an example is 2-methylpropan-2-ol as shown in Figure 7.1.5.

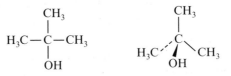

Figure 7.1.5

A tertiary alcohol

2-Methylpropan-2-ol

You might like to draw the other isomers of butanol, C_4H_9OH. There are two primary alcohols and one secondary alcohol. Can you name them? Which has two optical isomers? They have the same molecular formula but differ in their structural formulae, and also, in the case of the secondary alcohol, the arrangement of the atoms and groups in space.

SOME ETHERS ARE ISOMERIC WITH ALCOHOLS

$H_3C - O - CH_3$

Figure 7.1.6

Methoxymethane or dimethyl ether

Another structural possibility for compounds of the same general formula as alcohols, $C_nH_{2n+2}O$, is the ether functional group. Here, an oxygen atom is attached to two carbon atoms. There is no OH group present. For example, the first member of the ether series is when $n = 2$, CH_3OCH_3, dimethyl ether, or methoxymethane (Figure 7.1.6). The molecular formula is C_2H_6O, which could represent ethanol, C_2H_5OH, or an ether. These are functional group isomers since they contain different functional groups. The chemistry of ethers will not be discussed in any detail in this text. Ethoxyethane, $(C_2H_5)_2O$, known simply as 'ether', was used from 1841 until at least the 1950s as an anaesthetic for surgery.

7.1.3 Some Reactions of Alcohols

OXIDATION

Oxidation of alcohols occurs when the organic molecule reacts with oxygen or a reagent containing oxygen. Usually, two hydrogen atoms are taken away from the alcohol molecule. The OH is the alcohol 'functional' group, and the majority of the reactions of alcohols are to do with this group as it is more reactive than the C–H bonds of the rest of the molecule.

If you leave a bottle of wine (which contains alcohol) opened and exposed to the air, it will, after a while, start to smell (and taste) of sour apples. Leave it a little while longer and it will start to smell (and taste) of vinegar. This is because the wine contains ethanol (a primary alcohol) which reacts with and is oxidised by the oxygen in the air (shown as [O]) in the presence of micro-organisms such as *Mycoderma aceti*, as shown in Figure 7.1.7.

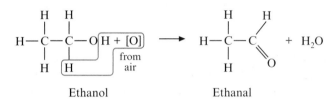

Ethanol   Ethanal

Figure 7.1.7
Primary alcohols are oxidised to aldehydes

In this reaction the alcohol is oxidised (each molecule loses two atoms of hydrogen), and the product is called an aldehyde. For ethanol, the aldehyde is called ethanal, 'ethan' because it has two carbon atoms and '-al' because it contains an aldehyde group (CHO) as part of the carbon chain. The names of all aldehydes end in 'al'. It is the ethanal that causes the wine to smell of sour apples when it begins to go sour in the air to form ethanoic acid.

'OK, Sir, I noticed that you were driving in a rather unusual manner, swaying from lane to lane.'

'Blow into this breathalyser please, Sir.' . . .

'Oh deary me', said the police sergeant sarcastically, 'What a surprise, it has gone green.'

'You must have been drinking rather a lot. Please leave your car here and accompany me to the police station.'

'Do you need help walking, Sir?'

The original breathalysers contained an orange oxidising agent [potassium dichromate(VI)] which when reacted with alcohol formed ethanal. The orange potassium dichromate(VI) was itself reduced to a green chromium compound. If your breath turned the breathalyser tube from orange to green, you must have had alcohol in your breath. The ethanal can then be further oxidised to ethanoic acid, by addition of an extra oxygen atom, as in Figure 7.1.8.

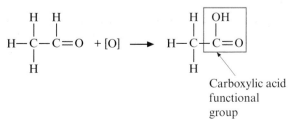

Carboxylic acid
functional
group

Figure 7.1.8
Ethanol can be further oxidised to ethanoic acid

thousands of years as a dilute solution by **fermentation**, which is the breakdown of molecules of sugars and starches in fruit, vegetables or grain – or even flowers – with water in the presence of the enzymes in yeast and the absence of air. The dilute solution can be distilled to concentrate the alcohol.

Fruit , vegetable or grain→ sugar → sucrose → glucose → alcohol
Each stage is catalysed by enzymes as in yeast.

The equation for the final stage is:

$$C_6H_{12}O_6(aq) \rightarrow 2C_2H_5OH(aq) + 2CO_2(g)$$
glucose   ethanol

TO THINK ABOUT . . .
(hint: look at the reactions of propene in Section 6.2.3)

How could propanol be synthesised from propene? Is propan-1-ol or propan-2-ol formed?

REDUCTION OF ALCOHOLS

Reduction in organic chemistry means the removal of oxygen from, or the addition of hydrogen to, a molecule. The removal of oxygen from the OH group and its replacement by hydrogen in an alcohol would produce an alkane. This cannot easily be done directly. It is usually done by making the alcohol into an alkene by dehydration, followed by addition of hydrogen under pressure with a catalyst; or by making the alcohol into a halogenoalkane, followed by reduction of the halogenoalkane with zinc and dilute hydrochloric acid.

$C_2H_5OH \rightarrow C_2H_6$ only achieved indirectly.
So: C_2H_5OH dehydrated to ethene C_2H_4.
C_2H_4 hydrogenated using hydrogen over a hot nickel catalyst $\rightarrow C_2H_6$

To what would propan-2-ol be reduced?

ESTER FORMATION

Alcohols react with carboxylic acids to form esters and water (Figure 7.1.12). An 'ester' contains the COOR group, where R can be any alkyl group, *e.g.* methyl, CH_3.

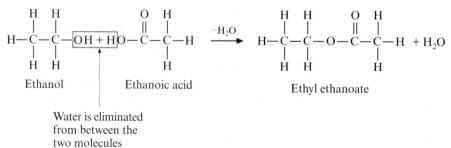

Figure 7.1.12
An esterification reaction

The ester is named from the alcohol and acid from which it is formed.

Esters are often fragrant, sweet-smelling compounds, and many plant flavours are esters. Some suitable ones are used in cooking and food preparations. Many of the exotic scents are either natural or synthetic esters.

Industrially, esters are used as solvents for paints, plastics and glues. They must be treated with care, as some if taken in large quantities can cause hallucinations, unconsciousness, permanent brain damage and death. The sweet smell of some glues has caused some, often younger people, to 'sniff' them, but solvent abuse has nothing but bad effects upon the brain, leading to brain cells being killed and eventually death for the abuser.

The equation for the formation of an ester by the reaction of an alcohol with a carboxylic acid (*e.g.* ethanoic acid), in the presence of some concentrated sulfuric acid to provide hydrogen ions as a catalyst, is:

$$CH_3COOH + C_2H_5OH + (H^+) \rightleftharpoons CH_3COOC_2H_5 + H_2O + (H^+)$$

The hydrogen ions are recycled (as befits a catalyst) and water is formed:

alcohol + acid $\rightleftharpoons$ ester + water

The opposite process to ester formation is the formation of an acid and an alcohol from the ester by hydrolysis (*i.e.* the action of water) in the presence of either an alkali or acid (*e.g.* the hydrolysis of ethyl ethanoate in Figure 7.1.13).

Fats are natural esters of propane-1,2,3-triol (glycerol). Soap used to be made by boiling animal fats with alkali; this is why the hydrolysis of esters used to be called 'saponification'.

> Pentyl ethanoate gives the flavour of pears. Methyl 2-hydroxybenzoate ('methyl salicylate') in low concentrations gives the flavour to peppermints; in higher concentrations it is used in pain-relieving muscle rubs such as 'Ralgex'.

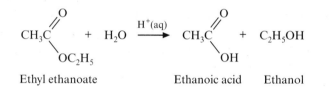

Ethyl ethanoate   Ethanoic acid   Ethanol

Figure 7.1.13
Hydrolysis is the reverse of esterification

In general, ester + water $\rightleftharpoons$ carboxylic acid + alcohol.

If an alkali is used for this hydrolysis, the salt of the carboxylic acid is formed and not the free acid. The salt then has to be treated with dilute sulfuric or hydrochloric acid in order to obtain the carboxylic acid.

SUBSTITUTION OF AN OH GROUP WITH Cl USING PCl₅ OR SOCl₂

The substitution of an OH group by a Cl group is achieved using phosphorous pentachloride, PCl_5, or sulfur dichloride oxide (or thionyl chloride), $SOCl_2$. This substitution reaction in primary alcohols requires these vigorous reagents, whereas the OH in the tertiary alcohols can be substituted by a Cl simply by using concentrated hydrochloric acid.

This ease of substitution provides a way in which primary, secondary and tertiary alcohols can be distinguished (Figure 7.1.14). In the Lucas test, an alcohol is shaken with zinc chloride in concentrated hydrochloric acid and allowed to stand. A tertiary alcohol reacts rapidly to give an opaque suspension of the insoluble chloroalkane. A secondary alcohol takes several minutes to become cloudy. A primary alcohol reacts very slowly, if at all.

Tertiary: $ROH + HCl \rightarrow RCl + H_2O$

Primary and Secondary need either PCl_5 or $SOCl_2$:

$$C_2H_5OH + PCl_5 \rightarrow C_2H_5Cl + POCl_3 + HCl$$
$$C_3H_7OH + SOCl_2 \rightarrow C_3H_7Cl + SO_2 + HCl$$

Figure 7.1.14
Substitution of an OH group

The alcohols are a very important group of compounds in organic chemistry because they have so many uses in making other organic compounds. Commercially, the alcohols are also important as solvents in the paint, printing, fabrics and dye industries.

Answers to diagnostic test

If you score less than 80%, then work through the text and re-test yourself at the end using this same test. If you still get a low score then re-work the unit at a later date.

1 The first three members of the alcohol series whose general formula is $C_nH_{2n+1}OH$, are methanol, ethanol and propanol.   **(3)**

2 The functional group of an alcohol is an OH, hydroxy group.   **(1)**

3 The alcohol, ethanol be made into the following:

a) an alkane. This is best made from an alcohol indirectly by first substituting the OH with a Cl using phosphorous pentachloride so forming the chloroalkane, then reducing this with hydrogen made from zinc and dilute acid; or by first making the alkene (see below) and hydrogenating the C=C with hydrogen gas using a hot nickel catalyst.   **(2)**

b) an alkene. Remove the OH and another H from an adjacent carbon atom (*i.e.* removal of H_2O) either with concentrated sulfuric acid or by passing the vapour of the alcohol over a hot surface, say hot aluminium oxide.   **(2)**

c) an aldehyde. Oxidation, by the removal of two hydrogen atoms, using acidified aqueous potassium dichromate(VI).   **(2)**

4 An alcohol can be made from propene in a series of reactions. Alcohols can be made *directly* by adding water (steam, under pressure plus a hot catalyst of phosphoric acid) to an alkene, but can also be made using a <u>series of reactions</u> by, *e.g.* propene + HBr → 2-bromopropane (in the absence of air) followed by hydrolysis with a hot dilute solution of sodium hydroxide.   **(2)**

5 a) The first member of the alcohol series to show position isomerism is propanol, which can be propan-1-ol or propan-2-ol.   **(1)**

b) The isomers and names are:

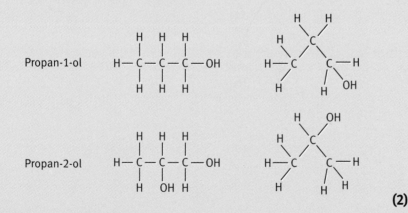

(2)

15 Marks   (80% = 12)

FURTHER QUESTIONS (ANSWERS GIVEN IN APPENDIX)

1 In the diagram below, match the following words with the letters A, B, C, D, E, F, G, H:

Oxidation, Reduction, Dehydration, Esterification, Hydrolysis, Hydration, Will not work directly in one stage.

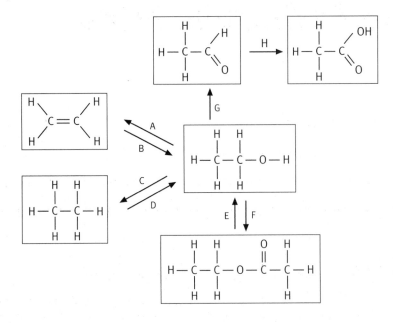

2 Write structures of C_3H_8O.

3 What will the secondary alcohol, propan-2-ol form on oxidation? (Hint, look at the chemistry of the carbonyl compounds first.)

4 What is the % composition by mass of ethanol? (Relative atomic masses: C, 12; H, 1; O, 16)

5 A compound has the following % composition: 60% carbon, 13.33% hydrogen, 26.66% oxygen. The compound has a relative molecular mass of 60. What is the molecular formula of the compound?

Unit 7.2
Carbonyl Compounds and Carboxylic Acids

Starter check

What you need to know before you start this unit.

You should understand the chemistry of the alcohols, the naming of organic compounds, and what is meant by sigma and pi bonds.

Aims **By the end of this unit you should be able to:**

- Provide an outline of the basic chemistry of the carbonyl compounds.
- Distinguish between the essential properties of aldehydes and ketones.
- Understand the chemistry of the carboxylic acids.
- Relate the chemistry of carbonyl compounds to other homologous series covered, including alcohols.

Diagnostic test

Try this test at the start of the unit. If you score more than 80%, then use this unit as a revision for yourself and scan through the text. If you score less than 80%, then work through the text and re-test yourself at the end by using this same test.

The answers are at the end of the unit.

1 Explain the difference between the arrangement of atoms around the carbonyl group in aldehydes, ketones and carboxylic acids. **(3)**
2 What is the general formula for the aldehyde series? **(1)**
3 Write the formula for the first aldehyde that has the same molecular formula as the first member of the ketone series. **(2)**
4 Give an example of an addition reaction of a carbonyl compound. **(2)**
5 How could a colourless liquid be chemically distinguished as either a ketone or an aldehyde, or as a carboxylic acid? **(4)**
6 Give reaction equations showing structural formulae for the formation of an aldehyde from a primary alcohol and the formation of a ketone from a secondary alcohol. **(4)**
7 Explain why methanoic acid can have some reducing properties, *e.g.* with Fehling's reagent. **(2)**

18 Marks   (80% = 14)

7.2.1 Carbonyl Compounds – General Formula $C_nH_{2n}O$

The carbonyl group, C=O, is found in a wide variety of compounds in a number of homologous series, namely aldehydes, ketones and carboxylic acids and their esters and amides (Figure 7.2.1). Compounds from these series exist in many natural products obtained from plants like rhubarb, citrus fruits, apples, and from the scents and flavours of fruits. Sugars also contain carbonyl groups. However, the behaviour of the C=O group in acids, esters and amides is greatly changed by the electronegative atom or atoms near it in the molecule. These behave very differently from a C=O in aldehydes and ketones, so when carbonyl compounds are mentioned in this unit it is **only** the aldehydes and ketones that are being referred to.

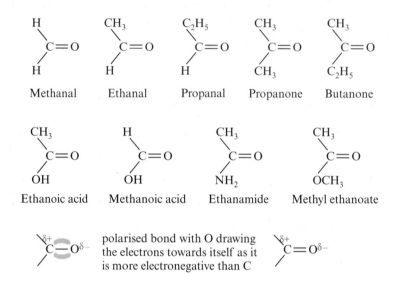

Figure 7.2.1
Homologous series containing the carbonyl group

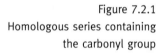

7.2.2 Some Properties of Aldehydes and Ketones

The term 'carbonyl compound' covers two distinct homologous series which are isomers of each other.

All aldehydes contain the H–C=O group, where the central carbon is attached to hydrogen or a further carbon, whereas the ketones contain the C=O group where the carbon atom is attached to two other carbon atoms.

The general formula of the aldehyde series is $C_nH_{2n+1}CHO$; the general formula for ketones is $(C_nH_{2n+1})_2CO$. From three carbon atoms onwards, therefore, pairs of isomers with the same molecular formula can exist; *e.g.* C_3H_6O could be an aldehyde, C_2H_5CHO, or a ketone, $(CH_3)_2CO$.

The boiling points given in Table 7.2.1 show that there is no hydrogen bonding between individual molecules of aldehydes and ketones, *i.e.* between the O of the CO bond and any hydrogen of an adjoining molecule, as is the case in carboxylic acids (compare with Figure 7.2.16). But there are intermolecular attractive forces between molecules due to the polarity of the C=O bonds, caused by the difference in electronegativity between C and O (see Figure 7.2.2) The intermolecular charge differences cause an interaction between different molecules due to their positive and negative charge attraction. An indication of the strength of these inter-attractive forces is shown by the boiling points of aldehydes and ketones (Figure 7.2.3). For a compound to boil, molecules of the substance must escape into the vapour state from the liquid. If there is attraction between each of the molecules in the liquid it is more difficult for a molecule to escape into the vapour. So a high boiling point shows that more energy has be put into the liquid for the molecule to break free. The higher the boiling point, the stronger the attractive forces between the molecules. Hydrogen bonding is stronger than the dipole attractive forces, and the weak van der Waals' forces, due just to electron fluctuations in covalent molecules and therefore dependant on molecular size, are the weakest of all. The data in Table 7.2.1 show some interesting trends for the molecules of similar mass. It shows the hydrogen bonding in alcohols has a greater effect than the dipole–dipole attraction in carbonyl compounds, which is in turn much stronger than the weak intermolecular forces which exist in the alkanes.

Table 7.2.1 *Relative molecular masses and boiling points*

RMM	Name	bp/K	RMM	Name	bp/K	Bonding
16	Methane	109	30	Ethane	184	No hydrogen bonding
32	Methanol	338	46	Ethanol	351	Hydrogen bonding
30	Methanal	252	44	Ethanal	294	Dipole attraction
46	Methanoic acid	374	60	Ethanoic acid	391	Hydrogen bonding
			59	Ethanamide	494	Hydrogen bonding
			58	Propanone	329	Dipole attraction
			60	Methyl methanoate	305	Dipole attraction

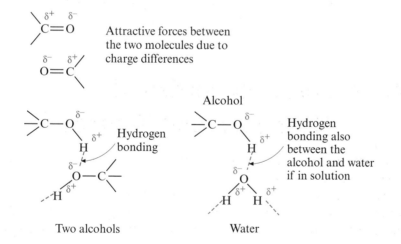

Figure 7.2.2
Dipole attraction (top) and hydrogen bonding (bottom)

Figure 7.2.3 shows how the boiling points of the first five members of each of the aldehyde and ketone series vary with relative molecular mass. Can you explain these graphs and why they are so similar?

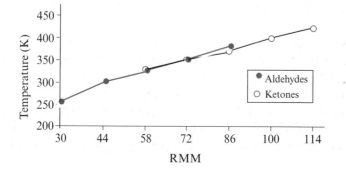

Figure 7.2.3
Boiling point *vs.* RMM for some simple aldehydes and ketones

7.2.3 Some Preparations of Aldehydes and Ketones

Aldehydes are made by oxidation from primary alcohols, and ketones are made by the oxidation of secondary alcohols (Figure 7.2.4).

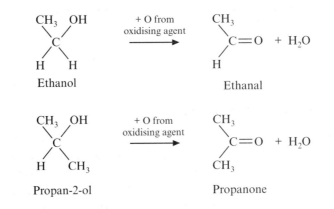

Figure 7.2.4
Aldehydes and ketones can be prepared from alcohols

The oxidising agent is usually acidified potassium dichromate solution.

When heating the mixture of the alcohol and the oxidising agent, the temperature can be set to allow the aldehyde to distil off and be collected; otherwise, further oxidation will occur and the corresponding carboxylic acid will be formed.

7.2.4 Reactions of Carbonyl Compounds

ADDITION REACTIONS

The presence of a double C=O bond in all aldehydes and ketones makes them open to addition reactions with those reagents strong enough to attach to the C=O bonds. The CO bond is polarised into a positive end (the C, δ^+) and a negative end (the O, δ^-), so reagents that can make use of the + and − electron density of the CO bond can attack these molecules. Reagents that contain an abundance of electrons will be attracted to the part of the CO bond that has an electron deficiency region (*i.e.* δ^+). Such reagents are said to be **nucleophiles**, *i.e.* those which seek areas of low electron density to deposit their negative electron abundance.

Atoms, ions and particles that have a deficiency of electrons are called **electrophiles**, and will seek out areas of high electron density, *i.e.* the δ^- of the O of the CO bond. Such reagents are quite numerous in chemistry and are very useful. The ones particularly useful for carbonyl compounds include the negatively polarised parts of: H–CN (hydrogen cyanide); H–NH$_2$ (ammonia and related compounds); and Na$^+$HSO$_3^-$ (sodium hydrogensulfite) (Figure 7.2.5).

In some addition compounds, if there is close proximity of an H atom and an OH group they can eliminate a molecule of water. The elimination of a small molecule, such as water, from an organic compound after an addition reaction is called a '**condensation**' reaction.

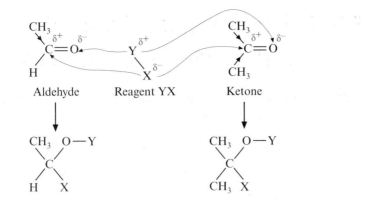

Figure 7.2.5
Nucleophiles seek areas of low electron density, and electrophiles, those of high electron density

Addition reactions with hydrogen cyanide, HCN

HCN adds across the C=O bond, with the $H^{\delta+}$ attacking the electronegative oxygen and the $CN^{\delta-}$ attacking the electron-deficient centre at the C atom of the CO (Figure 7.2.6). An industrial application for the HCN addition reaction is to make methyl methacrylate, the starting material for 'Perspex', an organic polymer and a non-breakable clear glass substitute used in car and aeroplane windscreens (Figure 7.2.7).

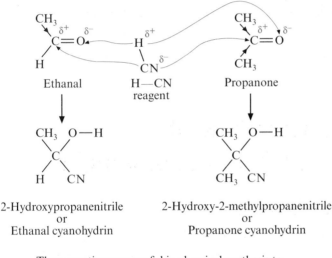

These reactions are useful in chemical synthesis to introduce an extra carbon atom into a molecule

Figure 7.2.6
HCN reacts *via* addition across the C=O bond

Addition reactions with sodium hydrogensulfite, NaHSO₃

In sodium hydrogensulfite the H^+ attacks the electron rich area at the O of the aldehyde or ketone C=O bond. The remainder of the compound adds to the C of the C=O (Figure 7.2.8). The resulting compound is crystalline.

'Condensation' reaction with hydrazine, NH₂NH₂. Addition followed by elimination

Addition of ammonia, NH_3, across the C=O double bond goes as expected (Figure 7.2.9). Hydrazine, H_2N-NH_2, adds similarly (Figure 7.2.10).

It can be seen that, after addition, one of the H atoms is in close proximity to the OH in the molecule; this results in the elimination of water. The resulting compound is called a hydrazone.

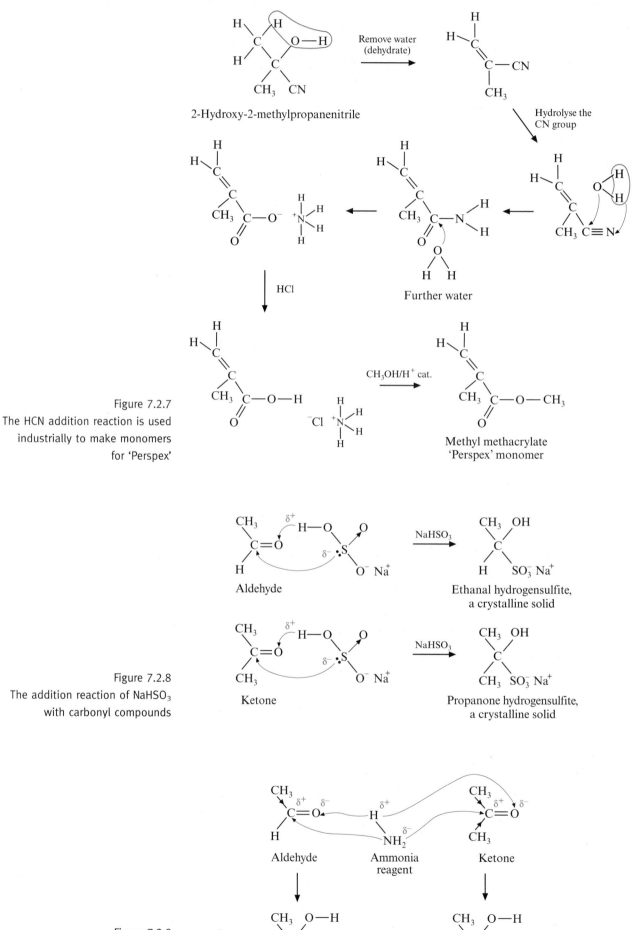

Figure 7.2.7
The HCN addition reaction is used industrially to make monomers for 'Perspex'

2-Hydroxy-2-methylpropanenitrile

Remove water (dehydrate)

Hydrolyse the CN group

Further water

HCl

CH_3OH/H^+ cat.

Methyl methacrylate 'Perspex' monomer

Figure 7.2.8
The addition reaction of $NaHSO_3$ with carbonyl compounds

Aldehyde

$NaHSO_3$

Ethanal hydrogensulfite, a crystalline solid

Ketone

$NaHSO_3$

Propanone hydrogensulfite, a crystalline solid

Figure 7.2.9
NH_3 adds across the C=O double bond

Aldehyde

Ammonia reagent

Ketone

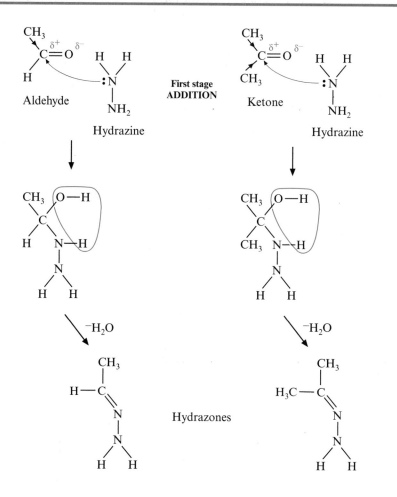

Figure 7.2.10
H_2N-NH_2 also reacts *via* addition; elimination of water then gives a hydrazone

A variation on this reaction is to use a substituted hydrazine, 2,4-dinitrophenylhydrazine (Figure 7.2.11). This is a heavier molecule and forms crystalline products with aldehydes and ketones more easily. It forms the corresponding 2,4-dinitrophenylhydrazone, and this reaction can be used to confirm the presence of a carbonyl compound.

'Condensation' reaction with hydroxylamine, NH₂OH

In a similar manner, the addition of hydroxylamine to both aldehydes and ketones results in an intermediate which eliminates water to give an oxime (Figure 7.2.12).

All of the addition compounds have characteristic melting points which should help identify the compound.

REDUCTION BY ADDITION OF HYDROGEN

Both aldehydes and ketones can be reduced by hydrogen and a hot cata-lyst, or by hydrogen liberated from sodium in alcohol or by any suitable reducing agent like $NaBH_4$ or $LiAlH_4$. The corresponding alcohol is formed (Figure 7.2.13).

OXIDATION

C–C bonds in any compound are not easily oxidised as a C=O bond in an aldehyde or ketone is already oxidised. In aldehydes, but not ketones, there is a C–H bond from the C which is also joined by a double bond to an O, and it is these C–H bonds that can be oxidised to a

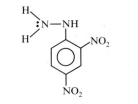

2,4-Dinitrophenylhydrazine or DNPH

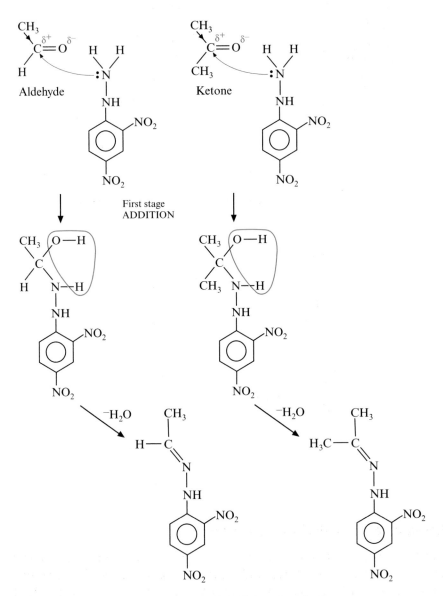

Figure 7.2.11
DNPH reacts *via* a condensation
reaction

2,4-Dinitrophenylhydrazones

C–O–H grouping to form a carboxylic acid group, $\overset{O}{\underset{OH}{C}}$, *e.g.* as in Figure 7.2.14.

So ketones cannot easily be oxidised but aldehydes can be. Indeed, aldehydes are so easily oxidised that they can be quite powerful reducing agents.

In the list given in Table 7.2.1, methanal, ethanal and propanal are oxidisable, whereas the propanone is not oxidised with ordinary common chemical reagents.

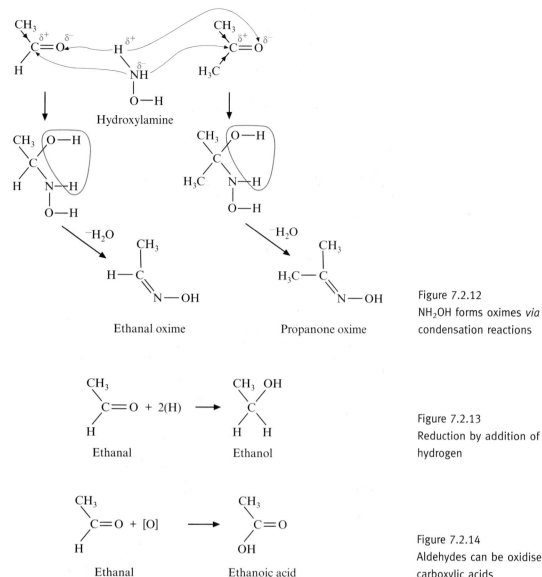

Figure 7.2.12
NH₂OH forms oximes *via* condensation reactions

Figure 7.2.13
Reduction by addition of hydrogen

Figure 7.2.14
Aldehydes can be oxidised to carboxylic acids

Chemical tests for aldehydes and ketones

Tollen's reagent, a solution of silver nitrate in ammonia, oxidises aldehydes to carboxylic acids, and the silver ions are reduced to a deposit of silver metal which often collects as a thin silver mirror on the inside of the test tube:

$$Ag^+ + e^- \rightarrow Ag \text{ metal}$$

Fehling's reagent contains complexed blue copper ions in the +2 oxidation state, and these are reduced when heated with aldehydes to a red precipitate of copper(I) oxide, Cu_2O. Ketones are not able to do these reactions.

Simple aldehydes and ketones will both burn if set alight in the air. The whole molecule is disintegrated and oxidised to carbon dioxide and water.

It might be of interest to prospective biology students to read further in the related chemistry of carbohydrates, as these contain carbonyl groups. See the extension topic at the end of this unit, Section 7.2.7.

7.2.5 Carboxylic Acids – General Formula $C_nH_{2n+1}COOH$

GENERAL PROPERTIES AND ACIDITY

Common carboxylic acids in the kitchen include dilute ethanoic acid (in vinegar); solid citric acid (2-hydroxypropane-1,2,3-tri-carboxylic acid) – the acid is also found in the juice of lemons and other citrus fruits; and tartaric acid (2,3-dihydroxy-butanedioic acid).

Carboxylic acids contain a carbonyl group, but it does *not* undergo the type of addition reactions that occur with the aldehydes and ketones. The carbonyl group in carboxylic acids, esters, amides or acyl chlorides has the electronegative atoms O, N or Cl next to the C=O, and these stop it from acting as a proper C=O group should (Figure 7.2.15).

The polarity of the bonds encourages ionisation to give the carboxylate anion. An alkyl group such as the methyl group feeds electron density towards the COOH group; this lessens the tendency to ionise, so ethanoic acid is weaker than methanoic acid.

Figure 7.2.15

The OH group has acidic properties due to the H tending to ionise. This is because the O atom is more electronegative than the H, and draws the bonding electrons towards itself and away from the H. In the presence of water, some of the acid molecules lose protons to the water:

$$CH_3COOH(aq) + H_2O(l) \rightleftharpoons CH_3COO^-(aq) + H_3O^+(aq)$$

It is these hydrated protons which give acid solutions their acidic properties. Carboxylic acids are usually only slightly ionised in solution – they are **weak** acids.

So long as there is no water around, carboxylic acids can hydrogen bond with each other. Ethanoic acid is able to form dimers (*i.e.* two molecules joining together) as shown in Figure 7.2.16.

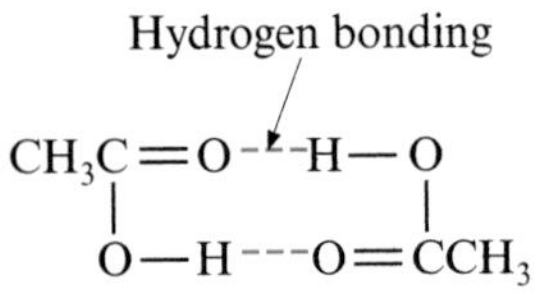

Hydrogen bonding

$$CH_3C=O^{-\text{-}}H-O$$
$$O-H\text{---}O=CCH_3$$

Figure 7.2.16

The first member of the series, methanoic acid, has a C–H bond attached to a CO, so it is similar in structure to the aldehydes. It can easily undergo further oxidation to carbonic acid, H_2CO_3, which falls apart to carbon dioxide and water. So, only methanoic acid in the carboxylic acid series, has reducing properties (Figure 7.2.17).

$$HCO_2H + (O) \rightarrow H_2CO_3 \rightarrow CO_2 + H_2O$$

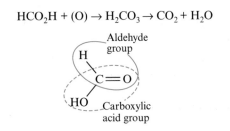

Figure 7.2.17
Only HCO_2H has some properties similar to those of aldehydes

The first two members of the carboxylic acid series are methanoic acid HCOOH (or the older name, formic acid) and ethanoic acid (or acetic acid), CH_3COOH (Figure 7.2.18).

CARBOXYLIC ACIDS WITH METALS

Carboxylic acids ionise when in water, some more easily than others, but all of those discussed above are weak acids. The equilibrium in the

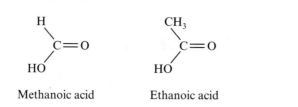

Methanoic acid   Ethanoic acid   Figure 7.2.18

ionisation of carboxylic acids is well to the left, and so only slight acidic properties result:

$$CH_3COOH(aq) \rightleftharpoons \text{ionises to } CH_3COO^-(aq) + H^+(aq)$$

Methanoic and ethanoic acids show the typical properties of acids, and will react with magnesium metal to give off hydrogen gas and liberate carbon dioxide gas from carbonates and hydrogen carbonates:

$$2CH_3COOH(aq) + Mg(s) \rightarrow Mg(CH_3COO)_2(aq) + H_2(g)$$

$$CH_3COOH(aq) + NaHCO_3(aq) \rightarrow CH_3COONa(aq) + CO_2(g) + H_2O(l)$$

ESTER FORMATION

Carboxylic acids react with alcohols to form esters, usually when heated under reflux, in the presence of a few drops of concentrated sulfuric acid. This reaction was discussed in the chemistry of the alcohols, Section 7.1.3.

OTHER REACTIONS

All except methanoic acid are not oxidised, except by burning to carbon dioxide and water.

They are not easily reduced, except by a very strong reducing agent, *e.g.* lithium tetrahydroaluminate(III), (LiAlH$_4$), more commonly known as lithium aluminium hydride. The product is the corresponding primary alcohol.

7.2.6 The Strength of a Carboxylic Acid

The strength of a carboxylic acid is determined by the degree of ionisation of the molecule.

Consider ethanoic acid (acetic acid, CH_3COOH): K_a is the equilibrium constant for the reaction, see Section 5.4.5:

$$CH_3COOH \rightleftharpoons CH_3COO^- + H^+$$

$$K_a = \frac{[CH_3COO^-][H^+]}{[CH_3COOH]}$$

$$pK_a = -\log_{10}(K_a)$$

The larger the value of K_a, the more the equilibrium favours the product. The 'p' in pK_a means the negative log, so the lower the pK_a value the stronger the acid.

You can see from Table 7.2.2 that the usual carboxylic acids get weaker as the carbon chain gets longer. Benzoic acid (an aromatic acid) is slightly stronger than ethanoic acid.

Table 7.2.2 *Strengths of some common carboxylic acids*

Name	Formula	K_a units/mol dm^{-3}	pK_a
Methanoic acid	HCOOH	1.6×10^{-4}	3.8
Ethanoic acid	CH_3COOH	1.7×10^{-5}	4.8
Propanoic acid	CH_3CH_2COOH	1.3×10^{-5}	5.0
Benzoic acid	C_6H_5COOH	6.3×10^{-5}	4.2

Answers to diagnostic test

If you score less than 80%, then work through the text and re-test yourself at the end by using this same test. If you still get a low score then re-work the topic at a later date.

1 The environment of the carbonyl group in aldehydes, ketones and carboxylic acids differs in the following ways:

- In aldehydes, A can be a further alkyl group, say methyl, and B **must** be a hydrogen atom.
- In ketones, both A and B are alkyl groups, say methyl groups.
- In carboxylic acids, A is hydrogen or an alkyl group, and B **must** be an OH group.   **(3)**

2 The general formula for an aldehyde is $C_nH_{2n+1}CHO$.   **(1)**

3 Ketones have a general formula $(C_nH_{2n+1})_2C=O$. Propanal and propanone are the first isomers of the carbonyl series to have the same molecular formula C_3H_6O. They are functional group isomers. (See Unit 6.4.)   **(2)**

4 An example of an addition reaction of a carbonyl compound is:

$$\underset{CH_3}{\overset{H}{\diagdown}}C{=}O \;+\; A{-}B \;\longrightarrow\; \underset{CH_3 \quad B}{\overset{H \qquad O{-}A}{\diagdown C \diagup}}$$

A–B can be H–CN, H–NH$_2$, Na$^+$HSO$_3^-$, *etc.*

(2)

5 A ketone, aldehyde or carboxylic acid can be distinguished from each other in the following ways.

First, test to see which one has acidic properties, *e.g.* **wet** blus litmus paper goes red with carboxylic acids.   **(1)**

If the remaining liquids are tested for a reducing agent with ammoniacal silver nitrate solution, the aldehyde should give a deposit of a silver mirror on the inside of the test tube: the other, the ketone, gives no deposit.   **(3)**

6 Equations for the formation of an aldehyde from a primary alcohol and a ketone from a secondary alcohol are:

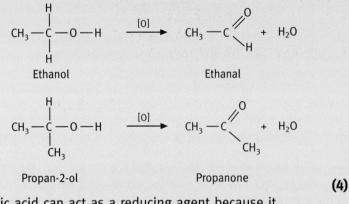

Ethanol   Ethanal

Propan-2-ol   Propanone   **(4)**

7 Methanoic acid can act as a reducing agent because it, uniquely for the carboxylic acids, contains a C–H bond which is easily oxidised.

Methanoic acid contains the CHO aldehyde reducing group and so gives some of the properties of aldehydes.

Methanoic acid   **(2)**

18 Marks (80% = 14)

FURTHER QUESTIONS (NO ANSWERS GIVEN)

1 Methanoic acid, HCOOH, is the first member of the carboxylic acid series; give the names and formulae of the next two members of the series.

2 Show what happens when ethanoic acid ionises, and how the structure of a carboxylic acid relates to the strength of the acid.

3 Does the C=O bond in the carboxylic acid show addition reactions similar to those of the aldehydes or ketones? Explain your reasoning.

4 Explain the meaning of the pK_a values for the strengths of the following acids: benzoic acid, 4.2; methanoic acid, 3.75; ethanoic acid, 4.76.

MORE ADVANCED QUESTIONS (NO ANSWERS GIVEN)

1 Explain why the mono-, di- and tri-chloroethanoic acids are progressively stronger acids than the original ethanoic acid. You should consider the inductive effect of the chloro group (electron attracting) on the O–H bond.

2 Will the chlorine atom on the benzene ring of 2-chlorobenzoic acid have a similar effect to the chlorine atoms in chloroethanoic acids?

3 Some carbohydrates give a silver mirror deposit with ammoniacal silver nitrate solution as aldehydes do. What does this tell you about the structure of these carbohydrates?

7.2.7 Extension – Aldehydes and Ketones in Sugars

'Sugar' is a simplistic name for a large series of carbohydrates. Carbohydrates contain C, H and O, and the H and O atoms are in the ratio 2:1. Sugars can contain aldehyde and ketone groups. The complete chemistry of these groups is beyond the scope of this book, but a few examples are shown in Figure 7.2.19.

Glucose

```
        CHO
   H──────OH
  HO──────H
   H──────OH
   H──────OH
       CH₂OH
```

This is an aldehyde

Fructose

```
       CH₂OH
        |
        C═O
  HO──────H
   H──────OH
   H──────OH
       CH₂OH
```

This is a ketone

These sugars can also exist in the ring form, *e.g.* glucose

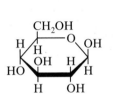

Figure 7.2.19
Some simple 'sugars'

The carbohydrates also have a lot of OH groups, which means a lot of hydrogen bonding. Note also how many possible optically active centres (asymmetric carbon atoms) there are in the molecules.

Sugars are sweet, and the chemistry is interesting but sometimes complex.

Unit 7.3
Amines

Starter check

What you need to know before you start this unit.

You should already know about chemical bonding and isomerism.

Aims **By the end of this unit you should understand:**

- Some of the chemistry of aliphatic amines.
- Some of the chemistry of aromatic amines.
- The differences between primary, secondary and tertiary amines.

Diagnostic test

Try this test at the start of the unit. If you score more than 80%, then use this unit as a revision for yourself and scan through the text. If you score less than 80%, then work through the text and re-test yourself at the end by using this same test.

The answers are at the end of the unit.

1 Explain the structural similarity between ammonia and the primary amine, methylamine. **(2)**
2 Give an example of a secondary amine. **(2)**
3 Show what happens when ethylamine dissolves in water. **(2)**
4 Explain the following; the pK_b value of ammonia is 4.75, of methylamine, 3.36, and of dimethylamine, 3.27. **(3)**
5 Explain why methylamine is a stronger base than phenylamine, $C_6H_5NH_2$. **(2)**
6 An amino acid contains both an amine group and a carboxylic acid group. Give an example of an amino acid. **(2)**
7 Phenylamine reacts with ethanoyl chloride, CH_3COCl. Give an equation showing the structures of both reactants and products. **(2)**

15 Marks   (80% = 12)

7.3.1 The Amines

SUBSTITUTION OF THE HYDROGEN ATOMS OF AMMONIA, NH₃

Amines are compounds with one, two or three of the hydrogen atoms of ammonia (NH_3) replaced by an alkyl or aromatic group.

It is possible to get isomers of amines with the same molecular formula which are primary, secondary and tertiary amines (Figure 7.3.1).

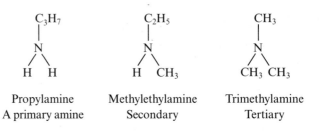

Figure 7.3.
Structural isomerism

Propylamine
A primary amine

Methylethylamine
Secondary

Trimethylamine
Tertiary

The amines are classed as **primary** if one H in ammonia has been replaced, **secondary** if two are replaced, and **tertiary** if all three H atoms are replaced (as in Table 7.3.1). Like ammonia, all of the amines are weakly alkaline in solution because they can accept protons from water molecules, so releasing hydroxide ions from the water into solution. They can also accept protons from acids to form salts (Figure 7.3.2).

Non-bonded pair of electrons

$$\ddot{N}H_3 + \overset{+}{H} \longrightarrow NH_4^+$$

Ammonia is a base and reacts with a proton to form an ammonium ion

Figure 7.3.2
Amines are basic compounds in solution

$$CH_3NH_2 + H_2O \rightleftharpoons CH_3NH_3^+ + OH^-$$

$$RNH_2 + H_2O \rightleftharpoons RNH_3^+ + OH^-$$

When all four H atoms of an ammonium ion are replaced by alkyl or aryl groups, compounds similar to ammonium salts are formed, as shown in Table 7.3.1.

Table 7.3.1

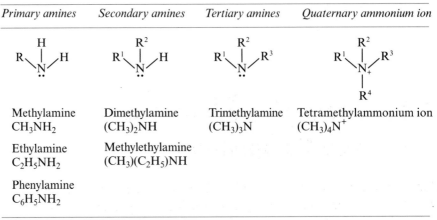

Primary amines	Secondary amines	Tertiary amines	Quaternary ammonium ion
Methylamine CH_3NH_2	Dimethylamine $(CH_3)_2NH$	Trimethylamine $(CH_3)_3N$	Tetramethylammonium ion $(CH_3)_4N^+$
Ethylamine $C_2H_5NH_2$	Methylethylamine $(CH_3)(C_2H_5)NH$		
Phenylamine $C_6H_5NH_2$			

Many of the amines have a fishy smell, and as early as 1851 trimethylamine was isolated by distillation of herring with lime and was called 'fish-gas'.

The smaller and earlier members of the amine series are gases, whereas the later ones are liquids or crystals.

7.3.2 Hydrogen Bonding in Amines

As can be seen from the structure shown in Figure 7.3.2 there is a 'lone pair' of electrons on the nitrogen atom in ammonia and amines, and these along with the electronegativity difference between N and H make the molecule polar. Amines are therefore often soluble in polar solvents, *e.g.* water. Hydrogen bonding occurs in ammonia and in primary and secondary amines. There are no hydrogens suitable for hydrogen bonding in tertiary amines (Figure 7.3.3).

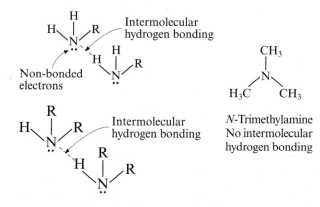

Figure 7.3.3
Hydrogen bonding does not occur between tertiary amines

In solution, hydrogen bonding occurs between the water molecules and ammonia or suitable amines (Figure 7.3.4). The hydrogen bonding is involved in the equilibrium between the amine and water on the one hand and the resultant ions on the other (see Section 7.3.5).

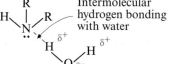

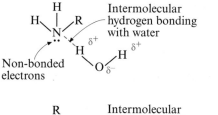

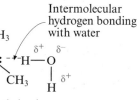

Figure 7.3.4
Hydrogen bonding can occur between all amines and water

7.3.3 Aromatic Amines and Azo Compounds

Some of the amines that contain aromatic groups are often used as starting points for making highly coloured 'azo' dyes (Figure 7.3.5). Some are used (controversially) as food dyes and some as dyes for cloth.

This 'diazotisation' reaction must be done at about 5 °C as the nitrous acid made from the sodium nitrite and dilute hydrochloric acid decomposes easily above about 10 °C.

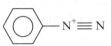
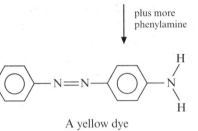

Phenylamine   + HNO₂

Nitrous acid (from
sodium nitrite
+ dilute acid)

Diazonium ion

plus more
phenylamine

A yellow dye

Figure 7.3.5
Azo dyes can be made *via*
'diazotisation' reactions

Another coloured dye is Ponceau red (E124), which is used in colouring tinned fruit.

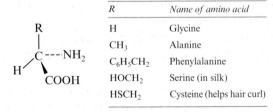

Ponceau red (E124)

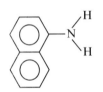

1-Aminonaphthalene
or
1-Naphthylamine

Figure 7.3.6

The starting amine for this compound was 1-aminonaphthalene (Figure 7.3.6). The ionic SO_3^- Na^+ groups on the dye makes it more water soluble, which is necessary for a food dye. Azo dyes have been linked with allergic reactions (*e.g.* asthma) in some people.

7.3.4 Amino Acids

Some compounds found in nature contain an amino group and a carboxylic acid group attached to the same carbon atom. Such compounds are called 'amino acids'; there are about 20 of them in Nature, and they are the building blocks of the proteins which are vital to you, not only as muscle builders but also as the enzymes which catalyse the reactions going on in every living cell.

Amino acids have the general formula $RCH(NH_2)COOH$. The R group can vary widely and some examples are given in Figure 7.3.7.

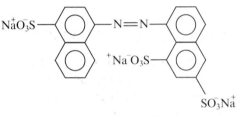

R	Name of amino acid
H	Glycine
CH_3	Alanine
$C_6H_5CH_2$	Phenylalanine
$HOCH_2$	Serine (in silk)
$HSCH_2$	Cysteine (helps hair curl)

Figure 7.3.7
Examples of some amino acids

For any amino acid except the first one, glycine (where R = H), the amines have an assymetric carbon atom in their molecules. Two optical isomers are possible (see Figure 7.3.8), as there are four different groups attached to the tetrahedral carbon atom. (See also Section 6.4.3 to remind yourself about optical activity.) This is important for all living things, as the natural amino acids (other than glycine) are all 'left-handed' and rotate polarised light anti-clockwise. So, for example, our digestive enzymes, containing amino acids in their proteins, cannot cope with proteins in food that are 'right-handed' (or which rotate polarised light clockwise). If you landed on a planet where amino acids and food proteins had all evolved with the opposite 'twist', you would die of starvation as your body would not be able to cope.

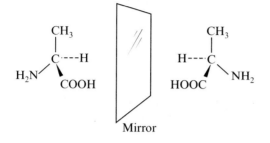

Figure 7.3.8
Optical isomerism in amino acids

CONDENSATION TO MAKE PEPTIDES AND PROTEINS

Amino acids can 'condense' together, losing a molecule of water each time a new bond is formed, to form chains of amino acids. Short or medium length chains give compounds called peptides; proteins consist of chains with perhaps hundreds or thousands of amino acid units joined together in them (Figure 7.3.9).

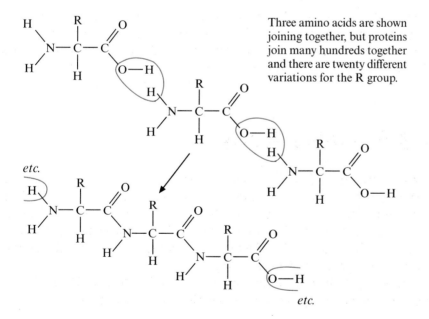

Three amino acids are shown joining together, but proteins join many hundreds together and there are twenty different variations for the R group.

Figure 7.3.9
Formation of proteins *via* condensation reactions of amino acids

The CONH group formed when two amino groups join together is called the **'peptide' link** or bond.

It must be pointed out that in a genuine peptide or protein, the peptide links all 'run' in the same direction, and are separated by only one C atom.

Use has been made of 'peptide link' formation when making 'artificial' molecules like Nylon. One type of Nylon can be made by the condensation of a dicarboxylic acid and a di-amine (Figure 7.3.10).

Notice that in Nylon the CONH links are in alternate directions and that there are several C atoms between them. So this is not a true peptide

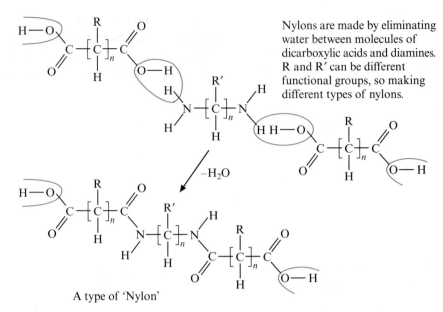

Nylons are made by eliminating water between molecules of dicarboxylic acids and diamines. R and R′ can be different functional groups, so making different types of nylons.

Figure 7.3.10
Formation of 'Nylon'

A type of 'Nylon'

link and our body enzymes cannot digest Nylon socks. Can you imagine wearing clothes which could be eaten when they were worn out? There's a research project for someone!

7.3.5 Base Strengths of Amines

The base strength of an amine can be defined using the following equations:

$$RNH_2 + H_2O \rightleftharpoons RNH_3^+ + OH^-$$

$e.g.$ $CH_3NH_2 + H_2O \rightleftharpoons CH_3NH_3^+ + OH^-$

The equilibrium constant is

$$K = \frac{[RNH_3^+][OH^-]}{[RNH_2][H_2O]}$$

In the same way as for K_a, because the concentration of water, $[H_2O]$, is large and changes very little it is incorporated into the value of the equilibrium constant which is re-named as K_b.

$$K[H_2O] = K_b = \frac{[RNH_3^+][OH^-]}{[RNH_2]} \quad \text{and} \quad pK_b = -\log_{10}(K_b)$$

The larger the value of K_b the stronger the base. The smaller the value of pK_b the stronger the base, because $K_b = -\log_{10}K_b$ (see Table 7.3.2) (see similar logic for the meaning of pH, Unit 5.5).

Table 7.3.2 *Base strengths of some common amines*

	K_b/mol dm^{-3}	pK_b
Ammonia	1.8×10^{-5}	4.75
Methylamine	4.4×10^{-4}	3.36
Dimethylamine	5.9×10^{-4}	3.27
Trimethylamine	5.8×10^{-5}	4.28
Phenylamine	4.2×10^{-10}	9.38

Methylamine is a stronger base than phenylamine, because in phenylamine the delocalised electon cloud of the benzene ring extends to overlap with the lone pair of electrons of the N. This makes it harder for the lone pair to donate to the a proton. So phenylamine is a weaker base (Figure 7.3.11).

Figure 7.3.11
Overlap between non-bonded electrons on N and the benzene electron cloud makes phenylamine a weaker base than methylamine

7.3.6 Some Further Reactions

The reactions of the amine group (NH_2), other than those due to its basicity, are all to do with the H atoms on the nitrogen being replaced by various organic groups. The names of the products are based on the longest carbon chain. Also, the acyl chlorides, ($RCOCl$), form compounds called substituted amides, which contain the $-CONH-$ link as in proteins and Nylon:

$$NH_3 + C_2H_5Cl \rightarrow C_2H_5NH_2 + HCl$$
ethylamine

$$CH_3NH_2 + C_2H_5Cl \rightarrow CH_3NHC_2H_5 + HCl$$
methyethylamine

$$CH_3NH_2 + CH_3COCl \rightarrow CH_3NHCOCH_3 + HCl$$
N-methylethanamide

$$\left(i.e. \quad CH_3 - \overset{\overset{O}{\|}}{C} - N \overset{H}{\underset{CH_3}{}} \right)$$

$$C_6H_5NH_2 + CH_3COCl \rightarrow C_6H_5NHCOCH_3 + HCl$$
N-phenylethanamide

(The N in the name indicates that the methyl- or phenyl group is attached to the nitrogen atom.)

REACTIONS WITH ACIDS

Amines are basic, as already shown, and so will react with acids to form salt-like compounds in a similar way to ammonia.

$$CH_3NH_2 + HCl \rightarrow CH_3NH_3{}^+Cl^-$$

$$NH_3 + HCl \rightarrow NH_4{}^+Cl^-$$

These reactions also occur in the amine group in amino acids and proteins.

$$H_2NCH_2COOH + H^+ \rightarrow {}^+H_3NCH_2COOH$$
glycine (or aminoethanoic acid)

The carboxylic acid group can, however, react with bases.

$$H_2NCH_2COOH + OH^- \rightarrow H_2NCH_2COO^- + H_2O$$

Note that, as amino acids can both accept and donate protons, they can undergo an internal acid–base reaction and end up with both positive and negative charges on the same molecule, *e.g.* $H_3N^+CH(R)COO^-$. This kind of ion is called a **zwitterion**, from the German meaning 'hermaphrodite' or 'hybrid'. It is this type of property that makes amino acids soluble in our stomach juices, and also ensures that they have high melting points.

We call compounds, like amino acids, that can act as acids or bases depending on their environment 'amphoteric' compounds. The body uses this dual property in the stomach, as the protein stomach juices and lining can react to 'mop' up any adverse effects of overeating or drinking (within reason). If the system is overloaded you might have to help a bit by taking a 'Rennie' or stomach powder to help restore the correct acidity to the stomach.

7.3.7 Summary of Reactions

1 Amines are substituted ammonia molecules and so have some similar properties. They are basic or alkaline compounds and their solutions have pHs above 7. Their base strength is expressed in terms of K_b or pK_b values.

2 Amines react with acids to form salts.

3 Ammonia can have its H atoms on the N progressively substituted by alkyl or aryl groups to form primary, secondary, tertiary and quaternary compounds.

4 Amino acids have both an amino group and a carboxylic acid attached to the same carbon atom in the molecule. This carbon is amphoteric, and can accept protons like a base, or donate protons, like an acid.

5 The H on the N of the amine group can also be substituted by other groups especially by acyl groups such as $COCH_3$, when attacked by ethanoyl chloride, CH_3COCl or ethanoic anhydride, $(CH_3CO)_2O$.

Answers to diagnostic test

If you score less than 80%, then work through the text and re-test yourself at the end by using this same test. If you still get a low score then re-work the topic at a later date.

1

$$H \overset{\overset{\displaystyle H}{\diagdown}}{\underset{\diagup}{\underset{\displaystyle H}{}}} N :$$

$$H \overset{\overset{\displaystyle H}{\diagdown}}{\underset{\diagup}{\underset{\displaystyle CH_3}{}}} N :$$

Non-bonded pair of electrons

Ammonia and methylamine have non-bonded electrons and are bases and so can pick up a proton to form ions. **(2)**

2 Secondary amines are any of those listed in Table 7.3.1. **(2)**

3 Ethylamine dissolves in water as shown in Figure 7.3.4. **(2)**

4 For the pK_b values; ammonia is 4.75, methylamine, 3.36 and dimethylamine 3.27, the lower the value of pK_b the stronger the base. Dimethylamine has two methyl groups on the N which feed electron density towards N by the inductive effect. This makes the electron cloud on the N stronger, so that the 'lone pair' of electrons on the N can be more easily donated to a proton. Ammonia has no methyl groups to help it enhance its electron pair donation (see Table 7.3.1). Methylamine is in between. **(3)**

5 Methylamine is a stronger base than phenylamine ($C_6H_5NH_2$). See answer to Question 4 above, but additionally in phenylamine the delocalised system of the benzene ring can extend to take in the 'lone pair' of the nitrogen, making it harder for the lone pair to be donated to a proton. **(2)**

6

Alanine

$H_2NCH(CH_3)COOH$

$$H_2N - \overset{\overset{\displaystyle CH_3}{|}}{\underset{\underset{\displaystyle H}{|}}{C}} - COOH$$

(2)

7 Phenylamine reacts with ethanoyl chloride, CH_3COCl:

$$C_6H_5NH_2 + CH_3COCl \longrightarrow C_6H_5NHCOCH_3 + HCl$$

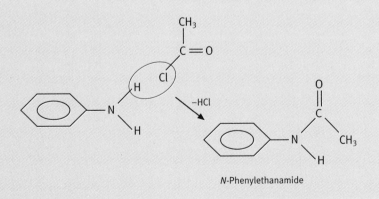

N-Phenylethanamide **(2)**

15 Marks **(80% = 12)**

FURTHER QUESTIONS (NO ANSWERS GIVEN)

1 Compare the basicity of dimethylamine with diphenylamine. Explain any differences.

2 Give an equation for the reaction of diphenylamine by ethanoyl chloride and name the product.

3 What class of amine is *N*-methyl-*N*-ethylphenylamine?

4 Dead fish when left to decay smell of methylamine. What does this tell you about the chemicals that are in fish?

5 Nylon was once made by condensing together two materials, 1,6-diaminohexane and a dicarboxylic acid, *e.g.* hexanedioic acid (with COOH groups at each end and four other CH_2 groups in between).

Draw out the structure of the compound formed when only two molecules of each of the two reactants have joined together.

Explain the similarities and differences between Nylon and proteins, in terms of how the units are linked together and the number of carbon atoms between the links.

Unit 7.4
Chemistry in Our Environment

Starter check

What you need to know before you start the contents of this unit.

This unit gives an overview of the effects of chemistry in our environment and draws upon the chemical ideas of many units of this course. It is probably best studied towards the end of the course.

Aims **By the end of this unit you should be able to:**

- Show understanding of some environmental issues.
- Appreciate the background chemistry of some of the reactions occurring in the environment.
- Relate areas of chemistry to everyday relevant topics of pollution and waste.

Diagnostic test

Try this test at the start of the unit. If you score more than 80%, then use this unit as a revision for yourself and scan through the text. If you score less than 80%, then work through the text and re-test yourself at the end by using this same test.

The answers are at the end of the unit.

1 What is meant by CFCs? Give the formula of one CFC.   **(2)**
2 What is the Montreal Protocol?   **(2)**
3 What is ozone? Why is it necessary to have ozone in the
   upper atmosphere?   **(2)**
4 Name two materials that damage the 'ozone layer'.   **(2)**
5 Give one equation to show how ozone is attacked by a
   chlorine atom produced from a CFC.   **(2)**
6 What is the gas caused by human activity which is causing
   an increase in the greenhouse effect? Explain its action.   **(2)**
7 Name two gases that are causes of acid rain.   **(2)**
8 What do catalytic converters on cars do?   **(2)**
9 How can acid rain affect aquatic plant life?   **(2)**
10 Give two examples of man-made materials that are recycled.   **(2)**

20 Marks   (80% = 16)

7.4.1 General Environmental Issues

The chemistry we use in our everyday life is generally beneficial to us; for example, processed foods, medicines, pharmaceuticals, scents, detergents, fibres and fabrics, plastics, processed metals, paints and wall coverings, dyestuffs, fuels, bricks and ceramics, improved food production by the use of fertilisers and insecticides, and many more. Society often forgets all of these, and it has become a fashionable thing to blame the scientists for all the pollution in our world. But we, the consumers, decide what we want for a 'better' lifestyle. We want more effective drugs, materials and food processing, *etc.*, but all development requires expensive and tested research.

We like to use refrigerators, but until recently many models used CFCs (chlorofluorocarbons) as coolants. Aerosols and spray cans also used CFCs as propellants. As CFCs can destroy the 'ozone layer' they have recently been replaced by more 'ozone friendly' propellants.

We expect foods that are free from toxic compounds and that have long shelf-lives before they go 'bad'. Chemists have worked to satisfy your needs. The foods are also sometimes irradiated with gamma rays to kill off any bacteria that could cause deterioration when stored for long periods.

Warm houses, heated by efficient fuels and insulated by double glazing and cavity wall insulation . . . thank the chemists for processing the natural gas and oils, and also for synthesising the polymers and plastics that make the window frames and foam cavity wall insulation.

'Have you got anything for my sore throat?', a person asks the pharmacist. Medicines are loaded with chemicals, often complex and usually in very small quantities. Many of the cures for diseases and ailments depend upon the work of chemists. All of the antibiotics are produced by fermentation, processing natural materials or by chemical synthesis. The intensive search for cures for cancer, AIDS, arthritis, migraines, and many other complaints involves thousands of chemists.

So why has the chemist got such a the bad image? Sometimes it has been because of the accidental misuse of beneficial materials, or of materials that were not realised to have side reactions. Glues and solvents come into this category. Because these substances can be abused by a small minority, should society say, 'we must do without them'? Do we stop driving cars because people are killed in car crashes? Most people stand far less chance of any bad effects from chemicals than you do from breathing in someone else's cigarette smoke, or from eating perishable food that has been left unrefrigerated or not cooked properly.

Here is a story which might start you thinking and debating the pros and cons of a discovery. Thomas Midgley was a first-class scientist. He used his understanding of chemical principles to solve two major industrial problems. He made the first effective additive for petrol which enabled car engines to use higher compression ratios, so making them more efficient. Midgley's answer was 'leaded' petrol in which he used tetraethyl lead as an additive to petrol.

The second of his great achievements was to find a working fluid for refrigerators and freezers which was non-toxic, non-corrosive and had the correct physical properties. He used a CFC.

His two major discoveries were welcomed by society by their massive use of more efficient engines and wide-scale use of refrigerators. Was it his fault that, in the 1990s, 'leaded' petrol and CFCs led to severe problems? It is easy *now* to see the bad effects. We must always try to apply our scientific knowledge with wisdom.

7.4.2 The Earth and its Atmosphere

For millions of years, the Earth, the oceans, the atmosphere, the weather, living systems and radiation from space have come to a state of equilibrium. When one set of external circumstances changes, then the remaining factors move to try to annul that change. Is this Le Chatelier's principle applying to nature? (see Unit 5.4). During the Ice Ages (caused by external factors) the remaining factors (including living systems) adjusted accordingly.

Human interference with the natural environment has contributed to destabilising the equilibrium. For example, the chopping down of forests equivalent to the size of Wales each week, with only minimal re-planting, and the building of great areas of radiation-reflective concrete and tarmac surfaces; the pouring of millions of tonnes of human waste into seas, and the depletion of marine life; the great increase in the numbers of cattle (which give out immense quantities of methane gas as waste, contributing to the greenhouse effect); the pollution of the air by car and industrial exhausts, *etc.*; all of these have affected the local and global environments. The vast Indonesian forest fires of September and October 1997, started by people clearing the ground for agriculture, had the effect of producing daytime smogs over much of South-east Asia. The particulates present in the atmosphere as a result of these fires could affect the global weather in areas all around the world for a number of months or years.

Can the Earth ever readjust?
Did it adjust in the past when huge volcanoes erupted?

The difference now is that mankind encroaches continuously upon the environment with no respite to allow the environment a chance to readjust. Any re-adjustments after natural catastrophes in the past might have taken 100–1000 years or even longer, but in our instant society we want to change everything in our lifetime, or especially tomorrow. We want a solution to the CFC problem and ozone depletion **now**, but the problem is so large that it will take a lifetime to solve, providing that all major nations want to solve it and comply with the banning of CFCs as recommended by the Montreal Protocol of the 1990s.

The world cannot revert to a self-adjusting environment because mankind puts ever increasing demands upon its resources. We take the natural rock, coal, air, oil and water continuously and process them into new materials, many of which will never degenerate to the original starting materials. The natural materials are lost for ever. If we want to provide a worthwhile world for the generations after us, then it is **we** who have to adjust. This generation is the first to take seriously the future needs of our planet. Not all countries see this as their main priority, and so some are working against progress in this battle.

The slogan should be 'Please leave this planet as you would wish to find it'. So a wise and proper use of chemistry is vital if we are going to do that.

7.4.3 The Atmosphere

There is more going on 'up there' in the atmosphere than we can imagine. There is a continuous in-pouring of particles and an outflow of infra-red radiation into space. Some of the radiation hitting the Earth is very harmful to living things, including us as fragile human beings. This includes high energy gamma rays, X-rays and ultra-violet light. There is also a stream of particles and high energy 'cosmic' rays, with such high energy they can pass straight through anything including yourself and the whole Earth. There is a layer of air – the atmosphere – that acts as a safety screen for us by removing much of the nastier radiation and by burning up meteorites. The composition of this layer is crucial for our continued existence.

Some of this radiation entering our atmosphere hits the atoms of oxygen and nitrogen, which are the major constituents of air. A whole collection of chemical reactions goes on, and they are all essential for maintaining the stable composition of the air both at high altitudes and at ground level. Here are some of the reactions.

FORMATION OF OZONE

O_2 + UV radiation (at about 30 km altitude) $\rightarrow$ 2O atoms, usually written as 2O*, where the * shows the splitting of the covalent bonds equally to give oxygen atoms with unpaired electrons. Then:

$O_2 + O^* \rightarrow O_3$ (ozone)

$O_3 + O^* \rightarrow 2O_2$

FORMATION OF IONS

O_2 + gamma rays and X-rays $[O_2^+ + e^-$ (an electron)]

N_2 + gamma rays and X-rays $[N_2^+ + e^-$ (an electron)]

N_2 + UV radiation at about 30 km altitude $\rightarrow 2N^*$ atoms

OXIDES OF NITROGEN

Nitrogen monoxide (NO) and nitrogen dioxide (NO_2) are almost unique in being small, common molecules which contain unpaired electrons.

$O^* + N^* \rightarrow NO$

$NO + O^* \rightarrow NO_2$

$NO + O_3 \rightarrow NO_2 + O_2$

$NO_2 + O_3 \rightarrow NO + 2O_2$

ANY WATER OR MOISTURE ALSO REACTS

$H_2O \rightarrow H^* + OH^*$

$H^* + O_3 \rightarrow OH^* + O_2$

$OH^* + O_3 \rightarrow HO_2^* + O_2$

$HO_2^* + O_3 \rightarrow OH^* + 2O_2$

The presence of any reactive hydrocarbon also affects the contents of the air by reacting with the various species in the upper atmosphere. Where R is a hydrocarbon group:

$$RH + OH* + O_2 \rightarrow RO_2* + H_2O$$

$$RO_2* + NO + O_2 \rightarrow \text{Carbonyl (R'C=O)} + NO_2 + HO_2*$$

$$HO_2* + NO \rightarrow NO_2 + OH*$$

There are many more reactions, also including a 'nuclear' one in which some of the nitrogen atoms of the upper atmosphere are hit by high energy cosmic rays, or particularly neutrons.

$$^{14}_{7}N + ^{1}_{0}n \rightarrow ^{14}_{6}C + ^{1}_{1}p$$

The radioactive carbon-14 isotope so formed is incorporated into carbon dioxide, and eventually by photosynthesis into plants and so into animals . . . including **you**. This is the basis of radioactive dating, as used by archaeologists (see also Section 1.4.4).

'All of these reactions continue at different rates and degrees of completion, but they successfully come to equilibrium to maintain the stable composition of the atmosphere. That is, until some pollutants come along and destroy one of the components, disturbing the equilibrium. It might take hundreds of years to readjust. Among the materials that can disturb this equilibrium are the CFCs (chloroflurocarbon compounds). Sherwood Rowland and Mario Molina discovered in 1974 that CFCs were unstable in the stratosphere (the upper atmosphere where ozone is abundant). These gases have been used in aerosols and refrigerators in the past partly because they are so stable under normal conditions. Over the years they have diffused into the upper atmosphere. For example, Freon-12 (CCl_2F_2) in the intense radiation of the upper atmosphere can give chlorine atoms which attack the ozone. In 1985, Joe Farman of the British Antarctic Survey detected a 'hole' in the ozone layer. Aircraft and satellites flying over the South Pole confirmed this and also detected the presence of the ClO radical. In 1987, the world scientific, industry and government representatives restricted the production of CFCs at a conference which produced the Montreal Protocol.

$$CF_2Cl_2 \xrightarrow{\text{UV radiation}} CF_2Cl* + Cl*$$

$$Cl* + O_3 \rightarrow ClO* + O_2$$

$$O_3 + ClO* \rightarrow Cl* + 2O_2$$

As can be seen from the equations, chlorine atoms are regenerated to attack more ozone molecules which break into oxygen, O_2. It is the 'ozone layer' that absorbs most of the high energy UV radiation that can be harmful to the skin by starting skin cancers, and, more importantly for the Earth, which can destroy the phytoplankton in the surface layer of the oceans – they are responsible for much of the photosynthesis, and hence are the starting point for food chains. You have heard weather forecasters telling you during the summer how you need to 'cover up' or wear protective skin creams to protect you from this harmful radiation. In Australia, and even in Britain, it has become quite a hazard, and people are warned not to stay out too long in the hot sun and to wear hats and sun creams.

Steps have been taken to replace all CFCs throughout the world, but some countries, owing to economic difficulties, are finding it hard to comply with the restrictions.

Some natural events also produce noxious gases and particles that can have a major effect on climate, *e.g.* large volcanic eruptions. When the volcano Tambora (Indonesia) erupted in 1815, the dust cloud in the upper atmosphere kept out sunlight from many places in the world, so that 1816 was known as 'the year without a summer'.

7.4.4 Photosynthesis

Many life-forms have developed to require a specific proportion of oxygen in the air. Animals in particular require oxygen for respiration, and they exhale carbon dioxide. Plants in sunlight take in carbon dioxide and moisture, and synthesise carbohydrates – sugar, starch and cellulose – from these. Ideally, photosynthesis and respiration are in equilibrium.

The equation for respiration based on glucose is:

$$C_6H_{12}O_6 + 6O_2 \rightarrow 6CO_2 + 6H_2O + \text{heat energy}$$

The equation for photosynthesis is:

$$6CO_2 + 6H_2O + \text{energy from sunlight} \rightarrow C_6H_{12}O_6 + 6O_2$$

It is essential, therefore, that green plants and the photosynthesising plankton in the oceans be conserved, and that trees and their leaves be allowed to work as efficiently as possible. Leaves should not be 'clogged up' by dirt and grime from pollution, or the trees be stripped of their leaves or made less efficient or smaller by acid rain and photochemical smog.

7.4.5 Fossil Fuels and Pollution

Fossil fuels when burned give out energy. We need fuels for transport, heating houses, industry, *etc.* In burning they produce a lot of carbon dioxide, which has been associated with the global greenhouse effects.

$$C(s) + O_2(g) \rightarrow CO_2(g)$$
Carbon as in coal

$$CH_4(g) + 2O_2(g) \rightarrow CO_2(g) + 2H_2O(l)$$
Methane, natural gas

$$2C_8H_{18}(g) + 25O_2(g) \rightarrow 16CO_2(g) + 18H_2O(l)$$
Octane as in petrol

Coal, oil, and natural gas are all called fossil fuels because they were formed from living things many millions of years ago. When we burn them we release the carbon in the form of the carbon dioxide which was taken out of the atmosphere all that time ago.

Since the Industrial Revolution about 300 years ago we have released the carbon dioxide taken out by probably a million years or so of photosynthesis.

Note also that because fossil fuels also contain some sulfur, burning them usually results in the formation of sulfur dioxide gas. This, when dissolved in falling rainwater, is one cause of 'acid rain'. Some modern power plants try to 'desulfurise' the gases before releasing them into the air.

7.4.6 What is the Greenhouse Effect?

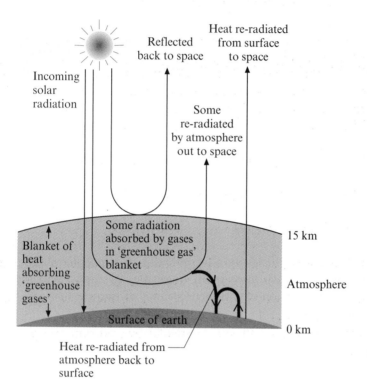

Figure 7.4.1

Schematic of the Greenhouse Effect

Much of the radiation from the Sun can get through the atmosphere, and hits the surface of the Earth. It heats up the soil and rocks, is absorbed by plants for photosynthesis, evaporates water from seas and wet ground, and helps to keep us warm.

Energy from the warm surface of the Earth radiates back into space, some of the energy as infra-red (IR) radiation. In the lower parts of the atmosphere near the Earth there is both carbon dioxide and water vapour, and it is these two gases which absorb the IR radiation. Their molecules therefore gain energy, which they pass to the surrounding air molecules by collisions. The air temperature therefore goes up. The atmosphere acts like double-glazed windows and lets light through but retains the energy as heat. The carbon dioxide, water vapour and other IR-absorbing gases thus control the temperature of our atmosphere. Without this blanket the Earth would have ice over much of its surface.

If too much CO_2, water vapour, methane *etc.*, are present, then too much heat is retained, and the effect is to warm up the lower atmosphere. The majority of atmospheric scientists now believe that global warming is occurring, but some believe that the Earth is able to readjust itself to accommodate these changes. (Le Chatelier again.)

The increased amount of CO_2 caused by vehicle exhaust fumes, and the presence of another heat-absorbing gas, methane, from cattle and paddy fields are beginning to cause concern. Some countries, *e.g.* Brazil, have tried to conserve hydrocarbon fuels by mixing petrol with alcohol, ethanol, made from fermenting their plentiful supply of sugar cane.

Burning alcohol as a fuel:

$$C_2H_5OH(l) + 3O_2(g) \rightarrow 2CO_2(g) + 3H_2O(l) + \text{energy given out}$$

The problem is that alcohol also makes CO_2 and water vapour when burned. However, at least it saves the hydrocarbon compounds for making other products such as polymers and plastics. Also, the sugar cane while growing takes in an equal amount of carbon dioxide from the atmosphere by photosynthesis.

Wood is another biomass fuel which, when growing, takes out as much carbon dioxide as it puts oxygen back by photosynthesis. In Britain the first fairly large wood-fuelled power station, using rapid-growing willow, began supplying electricity in late 1998.

7.4.7 Acid Rain

As we have seen, all fossil fuels contain sulfur, and sulfur dioxide is also produced when they are burnt, alongside the usual products of hydrocarbon combustion. This is one of the causes of acid rain pollution:

$$S(s) + O_2(g) \rightarrow SO_2(g)$$

Rain is always been slightly acidic due to atmospheric CO_2 dissolving in it, together with very small quantities of nitric acid from natural sources of oxidation of nitrogen in times of thunderstorms:

$$CO_2 + H_2O \rightleftharpoons H_2CO_3 \rightleftharpoons H^+ + HCO_3{}^-$$

$$N_2 + O_2 \rightarrow 2NO \xrightarrow{O_2\,(air)} NO_2 + H_2O \xrightarrow{O_2\,(air)} HNO_3 \rightarrow H^+ + NO_3^-$$

$$SO_2 + H_2O \rightarrow H_2SO_3 \xrightarrow{O_2\,(air)} H_2SO_4 \longrightarrow 2H^+ + SO_4^{2-}$$

In recent years the sulfur dioxide content of the air has risen, and so also have the nitrogen dioxide levels. This is due to increased use of fossil fuels and the fact that oxides of nitrogen are produced when some fuels and other materials burn in petrol engines. Catalytic converters on car exhausts help to reduce the amount of the oxides of nitrogen and carbon monoxide in the air, and so help to lower car 'emission pollution'.

Volcanoes also contribute to the concentration of particulates and of SO_2 and other gases in the air.

Power stations using fossil fuels are now expected to clean up all of their gases going into the air, and have invested heavily in apparatus and materials to do so. But large limestone quarries are required to produce the material to remove the acid gases, and these can sometimes leave large craters in sensitive landscape areas, *e.g.* the Yorkshire Dales. The product from using limestone in desulfurising plants is gypsum (calcium sulfate) and this has some value as a soil conditioner, and is used for blackboard chalk, plaster of Paris, plaster for walls and ceilings, *etc*. British Gypsum are extensively using 'power station' gypsum as a source of their 'raw material'; this, in turn, conserves the natural sources.

What happens to the carbon dioxide? Is it let out into the atmosphere to increase the concentration of 'greenhouse' gases?

7.4.8 Acid Rain and Plant Growth

The enriched acidity of the rain (mentioned above) means that when it falls on the ground it dissolves essential ions, including magnesium, out of the soil. Magnesium ions are essential for photosynthesis, and so plant growth is affected. Acid rain also dissolves some of the more toxic ions, such as aluminium, copper, lead and zinc, out of the soil. These metals under 'normal' conditions remain 'fixed' in the soil, but acid rain makes these toxic elements more available to plant life. These elements stunt the growth of plants. The water run-off from the soil into rivers and lakes means a build-up of toxic ions, and aquatic plants and fish are affected.

The Clean Air Acts of the various countries have helped to lower the acidity of the rain as well as removing smoke, so helping the natural restoration of the natural pH of lakes, rain, *etc.*

7.4.9 Solid Wastes and Other Materials

This section is a very brief description of the cause and effect of some pollutants in our environment which are there largely because of excessive consumption in richer countries.

Such pollutants include:

- liquid waste, detergents, sewage, oil waste;
- solid waste, metals (iron, mercury, copper compounds, *etc.*), plastics of various types, glass, processed paper, clothing, bedding, carpets, car batteries and other non-biodegradable household objects.

To do justice to all of these a major text would be needed, which is outside the scope of this project.

SOME EXAMPLES OF USING WASTE

- In Nottingham and some other towns, household waste is separated into combustible and non-combustible materials, and the combustible part used as a fuel to provide heat for buildings within the city.
- It has been estimated that a town of 500 000 generates 500 used tyres each week. Texaco has developed a plant to extract useful chemicals from them with the rest being used as a clean fuel for community use. A further use is to make crumbs of rubber for making safe surfaces underneath swings and in children's play areas.
- Many farmers use treated waste from sewage farms as a fertiliser. Some farmers use their own animal waste material to generate methane gas which can be used for domestic and farm heating supplies. Any solid material left is used as a fertiliser.
- Waste paper in vast amounts is recycled or burnt.
- Waste straw. The laws that prevent burning of this material have led many to apply their creativity to finding large-scale uses for it. One is to soak up accidental oil tanker spillages on coastlines. It can also be compacted with resin to make insulating boards, or compacted to make fuel blocks. Highly-insulated houses have been built using walls filled with straw bales.
- Bottle banks help to recycle waste glassware.

- Because so much electrical energy is used for making aluminium, it is good that more and more people now use skips to recycle aluminium drinks cans.

But, remember . . .when all of the energy involved in a recycling process has been calculated, recycling may not always be an energy-saving process, although it often is. It might, however, be environmentally beneficial. As a simple exercise for you to think about, consider the many tasks that are involved in returning ordinary plastic bags for recycling to a supermarket. Is the whole process hygienic, economic, material-saving, time-saving or socially acceptable? Which of the areas have the highest priority for you? Will they be the same for everyone?

7.4.10 Conclusion

This unit has probably led to more questions than real answers, and has also led you into more social awareness of chemical processes than other parts of this book. Without the understanding of chemistry which is developed in the earlier units of this text, you would not be as fully equipped to make a balanced judgement of the contrasting issues discussed. On the whole, the general public does not have an under-standing of chemical reactions, and so can be easily persuaded into making decisions based upon the knowledge of others. More informed decision-making can help both the chemical industry and society.

Knowing more chemistry can help you to become a more informed and a better citizen.

Answers to diagnostic test

If you score less than 80%, then work through the text and re-test yourself at the end by using this same test. If you still get a low score then re-work the topic at a later date.

1 CFCs are a group of compounds called **c**hloro**f**luoro**c**arbons, *e.g.* CCl_2F_2. **(2)**

2 The Montreal Protocol is a series of environmental guidelines restricting the production and use of CFCs which were drawn up at a conference in Montreal in 1987. **(2)**

3 Ozone is a molecule made up of three oxygen atoms, O_3. It is present in the upper atmosphere, where it is made by radiation breaking O_2 molecules into O atoms; some of the O atoms then attack the O_2 molecules to form as ozone molecules, O_3.

   The ozone layer acts as a screen by absorbing harmful ultra-violet radiation from the Sun, so protecting the Earth. **(2)**

4 CFCs drift up from the surface of our planet into the upper atmosphere destroy ozone molecules, *e.g.* CCl_2F_2 and $CHCl_2F$. **(2)**

5 Cl (atoms) from any source, including CFCs:

$$Cl^* + O_3 \rightarrow ClO^* + O_2.$$ **(2)**

6 The gas causing an increase in the greenhouse effect is carbon dioxide released by combustion of fossil fuels. It allows most of the Sun's radiation through to the Earth's surface; but, when the warm surface gives out infra-red radiation it stops this radiation from escaping into space. Hence, an increase in the concentration of CO_2 in the atmosphere makes the atmosphere a better insulating blanket than it already is. **(2)**

7 Sulfur dioxide and any oxide of nitrogen. **(2)**

8 Catalytic converters break down harmful oxides of nitrogen present in exhaust fumes to nitrogen and convert carbon monoxide to carbon dioxide, and so purify the gases let out to the atmosphere. **(2)**

9 Acid rain can fall on fields and soils and leach out aluminium and other metallic compounds which then enter the streams, rivers and lakes. These materials can kill off aquatic plants and so diminish food supplies for fish; also, many metal compounds in water are harmful to aquatic life. **(2)**

10 Paper, metals, and especially aluminium. It is more difficult to recycle plastics, but chemists are working on it. **(2)**

20 Marks   (80% = 16)

FURTHER QUESTIONS (NO ANSWERS GIVEN)

1 Give an account, as though you were writing a report for your local council, of how recycling facilities are essential for your community and for the country at large.

2 Write an account for your local paper of what is meant by each of the following. The article is to be read by lay-people with little or no scientific background.

- Acid rain;
- The greenhouse effect;
- The ozone layer.

Appendix
Selected Answers to
Further Questions

Unit 4.1

1 966 300; **2** 72.8; **3** 0.00482; **4** 2.02×10^6; **5** 1.11×10^{-6};
6 2.84×10^3; **7** 0.008; **8** 0.009; **9** 3.223; **10** 0.00769; **11** 0.0088;
12 3.22; **13** 5.985; **14** 1.862; **15** -2.317; **16** 6.305; **17** -5.955;
18 3.453; **19** -2.114; **20** -2.055; **21** 0.508; **22** 1.20×10^6;
23 1.016×10^5; **24** 1.486×10^{12}; **25** 1.08×10^{10};
26 $C = \dfrac{A-B}{n}$; **27** $X = \dfrac{50C}{AB}$;
28 a) 80 km h^{-1}; b) 7.5 s; c) Acceleration = 8 km h^{-1} s^{-1}.

Unit 4.2

1 $C_6H_{12}O_6$. Glucose is the same from any source.
2 $(1500 + 818) - 500 = 1818$ g
3 a) $H_2(g) + Cl_2(g) \rightarrow 2HCl(g)$
   b) $BaCO_3(s) \rightarrow BaO(s) + CO_2(g)$ already balanced
   c) $NaCl(aq) + AgNO_3(aq) \rightarrow NaNO_3(aq) + AgCl(s)$ already balanced
   d) $4Al(s) + 3O_2(g) \rightarrow 2Al_2O_3(s)$
   e) $Ca(OH)_2(s) + 2HNO_3(aq) \rightarrow 2H_2O(l) + Ca(NO_3)_2(aq)$
4 $MnSO_4(aq) + 2NaOH(aq) \rightarrow Mn(OH)_2(s) + Na_2SO_4(aq)$
5 $C_2H_5OH(l) + 3O_2(g) \rightarrow 2CO_2(g) + 3H_2O(l)$
6 $2HCl(aq) + Mg(OH)_2(s) \rightarrow MgCl_2 + 2H_2O$

Unit 4.3

1 33 g
2 93 g
3 6×10^{23}
4 a) $MgBr_2(aq) + 2AgNO_3(aq) \rightarrow 2AgBr(s) + Mg(NO_3)_2(aq)$
   b) 11.05 g
   c) 86.8%
5 a) Cu(I)Cl
   b) MnF_2
   c) CH_3O
6 a) C, 60%; H, 4.4%; O, 35.56%
   b) Ba, 78.3%; F, 21.7%
   c) C, 40.8%; H, 6.12%; N, 9.5%; O, 43.5%

Alphabetical List of Elements

Element	Symbol	Atomic number	Atomic mass†	Element	Symbol	Atomic number	Atomic mass†
Actinium	Ac	89	[227]	Mendelevium	Md	101	[258]
Aluminium	Al	13	27.0	Mercury	Hg	80	200.6
Americium	Am	95	[243]	Molybdenum	Mo	42	95.9
Antimony	Sb	51	121.7	Neodymium	Nd	60	144.2
Argon	Ar	18	40.0	Neon	Ne	10	20.2
Arsenic	As	33	74.9	Neptunium	Np	93	237.1
Astatine	At	85	[210]	Nickel	Ni	28	58.7
Barium	Ba	56	137.3	Niobium	Nb	41	92.9
Berkelium	Bk	97	[247]	Nitrogen	N	7	14.0
Beryllium	Be	4	9.0	Nobelium	No	102	[259]
Bismuth	Bi	83	209.0	Osmium	Os	76	190.2
Bohrium	Bh	107	[264]	Oxygen	O	8	16.0
Boron	B	5	10.8	Palladium	Pd	46	106.4
Bromine	Br	35	79.9	Phosphorus	P	15	31.0
Cadmium	Cd	48	112.4	Platinum	Pt	78	195.1
Caesium	Cs	55	132.9	Plutonium	Pu	94	[244]
Calcium	Ca	20	40.1	Polonium	Po	84	[209]
Californium	Cf	98	[251]	Potassium	K	19	39.1
Carbon	C	6	12.0	Praseodymium	Pr	59	140.9
Cerium	Ce	58	140.1	Promethium	Pm	61	[145]
Chlorine	Cl	17	35.5	Protactinium	Pa	91	231.0
Chromium	Cr	24	52.0	Radium	Ra	88	226.0
Cobalt	Co	27	58.9	Radon	Rn	86	[222]
Copper	Cu	29	63.5	Rhenium	Re	75	186.2
Curium	Cm	96	[247]	Rhodium	Rh	45	102.9
Dubnium	Db	105	[262]	Rubidium	Rb	37	85.5
Dysprosium	Dy	66	162.5	Ruthenium	Ru	44	101.1
Einsteinium	Es	99	[254]	Rutherfordium	Rf	104	261.1
Erbium	Er	68	167.3	Samarium	Sm	62	150.4
Europium	Eu	63	152.0	Scandium	Sc	21	45.0
Fermium	Fm	100	[257]	Seaborgium	Sg	106	263.1
Fluorine	F	9	19.0	Selenium	Se	34	79.0
Francium	Fr	87	[223]	Silicon	Si	14	28.1
Gadolinium	Gd	64	157.2	Silver	Ag	47	107.9
Gallium	Ga	31	69.7	Sodium	Na	11	23.0
Germanium	Ge	32	72.6	Strontium	Sr	38	87.6
Gold	Au	79	197.0	Sulfur	S	16	32.1
Hafnium	Hf	72	178.5	Tantalum	Ta	73	180.9
Hassium	Hs	108	[265]	Technetium	Tc	43	[99]
Helium	He	2	4.0	Tellurium	Te	52	127.6
Holmium	Ho	67	164.9	Terbium	Tb	65	158.9
Hydrogen	H	1	1.0	Thallium	Tl	81	204.4
Indium	In	49	114.8	Thorium	Th	90	232.0
Iodine	I	53	126.9	Thulium	Tm	69	168.9
Iridium	Ir	77	192.2	Tin	Sn	50	118.7
Iron	Fe	26	55.8	Titanium	Ti	22	47.9
Krypton	Kr	36	83.8	Tungsten	W	74	183.9
Lanthanum	La	57	138.9	Uranium	U	92	238.0
Lawrencium	Lr	103	[260]	Vanadium	V	23	50.9
Lead	Pb	82	207.2	Xenon	Xe	54	131.3
Lithium	Li	3	6.9	Ytterbium	Yb	70	173.0
Lutetium	Lu	71	175.0	Yttrium	Y	39	88.9
Magnesium	Mg	12	24.3	Zinc	Zn	30	65.4
Manganese	Mn	25	54.9	Zirconium	Zr	40	91.2
Meitnerium	Mt	109	[266]				

† [ ] indicates that element is radioactive.

Greek Symbols

Alpha	A	α
Beta	B	β
Gamma	Γ	γ
Delta	Δ	δ
Epsilon	E	ε
Zeta	Z	ζ
Eta	H	η
Theta	Θ	θ
Iota	I	ι
Kappa	K	κ
Lambda	Λ	λ
Mu	M	μ
Nu	N	ν
Xi	Ξ	ξ
Omicron	O	ο
Pi	Π	π
Rho	P	ρ
Sigma	Σ	σ
Tau	T	τ
Upsilon	Υ	υ
Phi	Φ	φ
Chi	X	χ
Psi	Ψ	ψ
Omega	Ω	ω

Subject Index

Absolute zero, 12, 13
Absorption spectra, 42
Acid
  and alkali, summary, 110, 260
  and calcium carbonate, 217–222
  dissociation constants, 260
  and living organisms, 262
  and metals, 107
  neutralisation, 109
  organic acids, carboxylic acids,
    326, 359
  pH calculations, 258
  pK_a, 261, 357
  rain, 382
  and sodium thiosulfate, 222
  strong, 261
  weak, 261
Activation energy (E_a), 211, 236
Addition reaction
  in alkenes, 297
  in carbonyl compounds, 352
  polymerisation, 299
Alcohols, 337ff
  homologous series, 338
  isomerism in, 325, 338
  primary, secondary and tertiary,
    339–340
  reactions of, 341–346
Aldehydes, 350
  comparison of properties with
    ketones, 350
  reactions of, 352ff
Alkali
  pH, 259
  pK_b, 370
  properties and general reactions
    of, 260
Alkanes
  general properties of, 285ff
  isomerism in, 321–330
Alkenes
  general reactions of, 285–289
  isomerism, 329–330
  polymerisation of, 299
Alpha particles and radioactivity,
  34–35
Amines, 366–368
  aromatic, 367
  hydrogen bonding in, 367
  primary, secondary and tertiary,
   366
Amino acids, 368
  as buffers, 267
Arenes, 310ff
Aromatic hydrocarbons, 311ff
  compounds of benzene, 312
  directing groups, 314
  isomerism, 326
  methylbenzene, 311

  substitution reactions, 313
Atmosphere of Earth, 377
Atomic mass units, 141, 150
Atomic number, 21
Atomic structure, 45, 46
Atoms, 20
  electrons in, 42
  and ions, 51
  nucleus of, 33
  and the periodic table, 30
  why and how they combine, 52
Avogadro's constant, 151
Azo dyes, 367

Balancing equations, 141
  rules for, 142
Batteries, electrical, 115,116
Becquerel, discovery of natural
  radioactivity, 34
Benzene, 310
Beta particles and radioactivity,
  34–35
Binary compounds, 94
Biosphere, 276–277
Boiling point elevation, 9, 194
Bond energies, 205, 229
Bonding
  covalent, 60–66
    giant structures, 68
    pi (π) bonds, 62
    shapes of orbitals and bonds,
     62
    sigma (σ) bonds, 62
  hydrogen, 72
  ionic, 53–58
    lattices, 68
  metallic structures, 69
  and properties, 67ff
    electrical conductivity, 71
    melting points, 70
    solubility in water, 71
Boyle's law for gases, 179
Breaking chemical bonds
  homolytically, free radicals, 301
Buckminsterfullerenes, 279
Buffer solution, 263–265
  amino acids as, 267
  carbonate-type, 266
  effect on body, 266
  phosphate-type, 266

Calcium carbonate and acids,
  217–222
Calculations, methods of approach,
  125
Calculator, use in calculations, 125
Cancelling out, 129
Carbonyl compounds, 350ff
  in aldehydes, 350

in carboxylic acids, 358
in ketones, 350
in sugars, 363
Carboxylic acids, 358ff
  isomers of, 326
  reactions of, 358
  strengths of, 359
Catalyst, effect on reaction rates, 234
  heterogeneous, 235
  homogeneous, 235
  'negative' or inhibitors, 236
CFCs, 372
Chadwick and the discovery of the neutron, 26
Changes of state, 5, 8
Charles's law for gases, 179
  combined gas law, 180
    calculations involving, 182
Chemical reactions of different types, 91ff
  acids and metals, 107
  decomposition, 97
  displacement, 106
  electricity, making, 114
  electrolysis, 101, 102
  exchange, 105
  fuels, 95
  heat energy of, 114
  ion exchange, 108
  neutralisation, 109
  organic synthesis, 95
  redox, 116
  synthesis, 93, 94
  thermal decomposition, 99
Classification of the elements
  Dmitri Mendeleev, 29
  Dobereiner, 29
  history of, 29
  Lothar Meyer, 29
  modern periodic table, 28
  Newlands, 29, 30
Colligative properties, 194
Collisions and reaction rates, 204
Colloidal dispersion, 196
Colloids, 196
Combined gas equation, 180
  calculations involving it, 182
Combining and synthesis, 93
Composition, percentage, 157
Compound, 19
Concentrated acids, 168
  their dilution, 169
Concentration
  effect on reaction rate, 205, 231
  of solutions, mole concentrations, 164
  units of, 164
    parts per billion, 170
    parts per million, 170
    percent by mass, 170
    percent by volume, 170
Condensation
  as an organic reaction for carbonyl compounds, 352–357
  as a physical process, 6, 7

Condensation polymerisation, 302
Conductivity, 71
Configuration of electrons in atoms, 45, 46
Contact Process (manufacture of sulfuric acid), 248–249
Covalent bonding, principles and examples, 59–66
Cracking of long chain hydrocarbons, 287–288
Crystal lattices, 68

Decomposition reaction, 80, 97
  thermal decomposition, 80, 99, 243
Dehydration of alcohols, 340
Density of a gas, 184
Diatomic molecules of gases and bonding in, 60–61
Diffusion
  in gases, 10, 175
  in liquids, 11
Diluting solutions, 167
Displacement reaction, 79, 106
Dissociation, thermal, 243
Dissolving, 11
Dobereiner triads, 28
Dynamic equilibrium, 251

Electrical conductivity of compounds, 71
Electricity from a chemical reaction, 114, 115
  Voltaic pile, 116
Electrolysis
  of aqueous solutions, 102
  of molten compounds, 101
Electronic configuration, 21
  and reactivity, 75
Electrons
  in atoms, 42
  gain of, 77
  loss of, 76
Electrophiles, 352
Elementary particles, 26
Elements, 18
  Ancient Greek meaning of, 18
  classification of, 28, 29
  joining together, 94
  transmutation of, 35
Empirical formula, 159
Endothermic and exothermic changes, 114, 207, 208
Energy, activation (E_a), 211, 236
Energy changes in reactions, 203, 206
  bond, 205
  endothermic, 208
  exothermic, 208
Enthalpy, 208, 210
Enzymes
  inhibitors for rates of reactions, 239
  uses of, 238
Equations, mathematical rearrangement of, 127

Equilibrium
  chemical, 244
  effect of catalysts on, 247
  effect of temperature on, 247
  factors affecting, 245
  industrial importance of, 247
  law of, 251
  physical, 244
  removal of products, 246
Esters
  formation of, 344, 359
  isomers of, 326
Ether isomers of alcohols, 325, 340
Evaporation, 6, 9
Exchange reactions
  acids and metals, 107
  displacement, 106
  ion exchange, 108
  neutralisation of acids, 109
  radical exchange, 106
Exponential graphs, 131
Exponential numbers and powers of
  10, 122, 123

Fast reactions
  methane and chlorine, mechanism,
    306
Formula mass, 152
Fractional distillation, 288
Free radicals, 301
Freezing point depression, 194
Fuels, 95
  fossil fuels and pollution, 380
Functional group isomerism, 325
Functional groups
  in alcohols, 337ff
  in aldehydes, 350ff
  in amines, 366ff
  in carboxylic acids, 358ff
  in ketones, 350ff

Gas constant (R), 182
Gases
  collisions and kinetic theory of
    gases, 178
  density of, 184
  ideal gas, 178, 182
  kinetic theory of, 178
  molar volume, 177
  partial pressure, 184
  pressure, 176
  standard temperature and
    pressure, 177
  temperature, 177
  volume, 177
Gas laws
  Boyle's Law, 179
  Charles's Law, 179
  Combined Gas Law, 180
Geometric (cis/trans) isomerism,
  329ff
Giant covalent structures, 68
Giant ionic crystal structures, 68
Giant metallic structures, 69
Graphite, 276
Graphs, 129

exponential, 131
gradient or slope of, 130
Greenhouse effect, 381

Haber Process (manufacture of
  ammonia), 248
Half-life and radioactivity, 36, 37
'Heat'
  energy, 114
  of reaction (H), 207, 210
Heterogeneous catalysts, 235
Homogeneous catalysts, 235
Homologous series, 276
Hydration, 189, 343
Hydrocarbons
  alkanes, 285ff
  alkenes, 285ff
  arenes, 309ff
Hydrogen bonding, 72
  in alcohols, 339
  in amines, 367
Hydroxonium ions, 260

Ideal gas, 178
Industrial manufacture of
  nitric acid
    Haber Process, 248
    Ostwald Process, 250
  sulfuric acid, Contact Process, 248,
    249
Inert gas structure, 52
Inhibitor
  for catalysts, 236
  of enzymes, 239
Initiator for polymerisation, 301
Ion exchange, 108
Ionic bonding
  principles of, 53–55
  properties of, 53, 54
Isomerism
  aromatic, 326
  functional group, 325
  general classes of, 321ff
  geometric or cis/trans, 329
  optical, 327
  positional, 324
  structural, 321
Isotopes, 27

K_a (acid dissociation constant), 261,
  359
K_c (equilibrium constant), 251
K_w (ionic product of water), 257
Kelvin, temperature, 177
Ketones, 350ff
Kinetic energy in reactions, 230
Kinetics of reactions, 239
Kinetic theory of gases, 178

Lattices of crystals, 68
Lavoisier, Antoine, 99
Law of conservation of energy, 208
Law of conservation of mass, 139
Law of constant composition, 139
Law of mass action of Guldberg and
  Waage, 250

Le Chatalier's principle, 246
Limestone
    with acids, reaction rates,
        217–222
    decomposition, 100
Limiting quantities, 156
Logarithms, 126, 135
Lothar Meyer, *see* Classification of
        the elements

Markownikov rule for addition to
        alkenes, 299
Mass
    molar, 152
    number, 26, 27
    relative atomic mass (RAM), 27,
        140, 151
    relative formula mass (RFM), 152
Mathematical concepts, 121–136
Melting point, 6
    of covalent and ionic compounds,
        70
Mendeleev, Dmitri, *see* Periodic table
Meyer, Lothar, *see* Classification of
        the elements and Periodic
        table
Molar excess, 156
    limiting quantities, 156
Molarity, calculations involving,
        164
Molecules and sharing electrons,
        59ff
Moles, calculations of, 151
Monitoring chemical change, 223

Naming of organic compounds,
        system, 289
Neutralisation, 109
Neutrons, 26
Newlands's octaves, *see*
        Classification of the elements
Nitric acid manufacture
    Haber Process, 248
    Ostwald Process, 250
Nucleophile, 350
Nucleus of an atom, 33
Numbers
    atomic, 21
    exponential, 122, 123
    logs of, 126
    powers of, 122
    significant figures, 127
    standard form, 122–124
    using calculators, 125
Nylon, 302, 369–370

Orbitals of electrons in atoms,
        44–46
    pi-clouds, 62
    shapes in covalent bonds, sigma,
        62
Organic synthesis, 95
Osmotic pressure, 194
Ostwald process (manufacture of
        nitric acid), 250
Oxidation, 83–92

Oxidation number, 83–92
    and formulae, 85
    and group, 84
    of multi-atom ions, 88
    of transition metals, 85
Ozone, 373

Partial pressure, 184
Particles
    atoms, 20
    electrons, 20
    of materials, 7
    protons and neutrons, 20
    size and effect on rates of
        reaction, 232–233
Parts per billion (ppb), 169, 170
Parts per million (ppm), 169, 170
Percentage composition, 157
Percentage by volume, 171
Percentage yield, 155
Periodic table
    Dobereiner, 28
    history of discovery, 29
    Lothar Meyer, 29
    Mendeleev, 29
    modern classification of, 28
    Newlands, 29, 30
    periodicity
        of atomic masses, 29
        of atomic number, 28
        of atomic radii, 28
    trends in, 78
pH
    of acids, 258
    of alkalis, 259
    calculations, 258
    uses, 262
Photosynthesis, 380
Pi bonding, 280, 291
pK_a; *see also* Acid, dissociation
        constant, 261, 359
pK_b, 370
Plant growth and acid rain, 383
Polarity of covalent bonds, 60, 61
Pollution, 376
    fossil fuel, 380
Polymerisation
    addition, 300
    condensation, 302
Polymers/plastics, 291
Positional isomerism, 324
Powers of 10, 122
    multiplying them, 125
Pressure
    effect on gaseous reactions, 246,
        248
    of gases, 11, 12
    gas laws, Boyle's and Charles's,
        179
    partial, 184
Priestley, Joseph, 99
Properties
    chemical, 5
    of covalent and ionic compounds,
        67–74
    physical, 5

Proteins and peptides, 369
Proton, 26
Pure, its scientific meaning, 18

Quantum numbers, 43, 44
    azimuthal, 44
    magnetic, 44

Radioactivity
    detection of, 34
    discovery of, 34
    types of radiation, 34
    uses of, 38
Radical exchange, 306
Rates of reaction
    effect of catalysts and enzymes
        on, 234, 238
    effect of concentration on, 205,
        231
    effect of particle size on, 232–233
    effect of pressure on, 232
    effect of temperature on, 230
    measurements of, 217
Reacting quantities and the mole,
    153–154
Reactivity series
    for metals, 78
    for non-metals, 81
    as a predictor for reactions, 79–81
Rearranging equations, 127
Recrystallisation, 192
Redox reactions, 83–92, 116
Reduction, 83–92
    of alcohols, 342
Relative atomic mass (RAM), 27
Relative formula mass (RFM), 152
Relative molecular mass (RMM),
    152
Reversible reactions, 243
'Rounding up' numbers, 127

Saturated solution, 168
Sharing electrons, 59
Sigma bonds, 280
Significant figures, 127
    'rounding up', 127
Solid waste, 383
Solubility of compounds in water,
    71, 190–193
Solutions
    colligative properties of, 193
        depression of freezing point,
            194
        elevation of boiling point, 194
        osmotic pressure, 194
        vapour pressure, 193
    colloidal suspensions, 196
    immiscible solvents, 189
    insoluble, 188
    miscible solvents, 189

partially miscible, 189
saturated, 168
solutes in, 188
solvation and hydration in, 189
solvents for, 188
suspensions, 188
types of, 190
Solvation, 189
Spectroscopy
    absorption, 43, 44
    evidence for bonding from, 43
Standard conditions of temperature
    and pressure (STP), 177
Standard form, for exponential
    numbers, 122–124
    calculations involving, 124
States of matter, 5, 6
    changes of state, 8
    gas, 6
    liquid, 6, 8
    solid, 6, 8
Stereo-isomerism, 327
Strengths of solutions
    concentration units, 164
    diluting solutions, 167
    saturated solutions, 168
Structural isomerism, 321
Study guide, xix
Substitution reactions
    in alcohols, 345
    in arenes, 313
Sugars, 363
Symbols for elements, 26
Synthesis, 94, 95

Temperature effect on reactions,
    205, 217, 229
Termination of polymerisation, 301
Thermal decomposition, 80, 243
Transition metals, 85
Transmutation of elements, 35
Types of chemical reaction, 91

Units from graphs, 132

Vapour pressure, 176, 184

Waste material and its use, 379
Water and bonding in, 20, 61, 62
    hydrogen bonding, 72
Wavelengths of light, 42

Yield
    actual, 155
    calculation of, 155
    percentage, 155
    theoretical, 155

Zero, absolute, 12, 13
Zwitterions in amino acids, 262, 372

DISCARDED

DATE DUE			
JUL 29 2010			
AUG 16 2010			
SEP 25 2013			
GAYLORD			PRINTED IN U.S.A.